Bioinformatics

Volker Sperschneider

Bioinformatics

Problem Solving Paradigms

With contributions by Jana Sperschneider
and Lena Scheubert

 Springer

Volker Sperschneider
University of Osnabrück
Department of Mathematics
and Computer Science
Albrechtstr. 28
49076 Osnabrück
Germany
informatik.uni-osnabrueck.de

Contributors:

Lena Scheubert
University of Osnabrück
Department of Mathematics
and Computer Science
Albrechtstr. 28
49076 Osnabrück
Germany

Jana Sperschneider
The University of Western Australia
School of Computer Science
and Software Engineering
35 Stirling Highway
Perth, WA 6009
Australia

ISBN 978-3-642-09726-3   e-ISBN 978-3-540-78506-4

DOI 10.1007/978-3-540-78506-4

Cover design: KuenkelLopka Werbeagentur, Heidelberg

Printed on acid-free paper

9 8 7 6 5 4 3 2 1

springer.com

Dedicated to my family

Preface

What is this book good for?

Imagine you are a computer scientist working in the bioinformatics area. Probably you will be a member of a highly interdisciplinary team consisting of biologists, chemists, mathematicians, computer scientists ranging from programmers to algorithm engineers, and eventually people from various further fields. A major problem within such interdisciplinary teams is always to find some common language, and, for each member of some discipline, to have profound knowledge of what are the notions, basic concepts and goals of the other participating disciplines, as well as of what they can contribute to the solution of ones own problems. This does, of course, not mean that a computer scientist should do the job of the biologist. Nevertheless, a computer scientist should be able to understand what a biologist deals with. On the other hand, the biologist should not do the computer scientists job, but should know what computer science and algorithm engineering might contribute to the solution of her/his problems, and also how problems should be stated in order for the computer scientist to understand them.

This book primarily aims to show the potential that algorithm engineering offers for the solution of core bioinformatics problems. In this sense, it is oriented both towards biologists indicating them what computer science might contribute to the solution of application problems, as well as to mathematicians and algorithm designers teaching them a couple of fundamental paradigms for the analysis of the complexity of problems and, following this analysis, for the design of optimal solution algorithms. Thus the goal of the book is neither to present a more or less complete survey of bioinformatics themes, nor to do programming or introduce the usage of bioinformatics tools or bioinformatics databases. It is a book on *fundamental algorithm design principles* that are over and over again applied in bioinformatics, with a clear formal presentation what an algorithm engineer should know about each of these principles (omitting lots of things that are more intended for a theoretically complete treatment of themes), and lots of case studies illustrating these principles. The selection of case studies covers many of the milestones in the development of bioinformatics, such as genome sequencing, string alignment, suffix tree usage, and many more, but also some more special themes that are usually omitted in textbooks as "being too complicated"[1], for example, the quite complicated derivation of the formula computing the directed reversal distance between two signed genomes, as well as a few themes that recently attracted attention, for example, analysis of molecular networks or prediction of RNA pseudoknots.

Book Overview

Chapter 1 starts with an informal presentation of *core bioinformatics problems*. From this, a computer scientist or mathematician can draw some first impression of what sort of problems bioinformatics deals with. Most of the presented problems are milestones of the development of bioinformatics, and a few are highly relevant problems that have attracted attention over the past few years.

Chapter 2 sheds some light on what sort of *formal algorithmic problems* evolve from the biological problems presented in Chap. 1. Here, a biologist may learn about the basic notions and concepts a computer scientist uses. For some of the presented algorithmic problems, solutions are given already in this chapter or at least some preparations for their solution in later chapters.

Chapter 3 presents one of the most useful algorithm design techniques in bioinformatics (but also in many other applications areas), *dynamic programming* which is based on the fundamental divide-and-conquer paradigm of recursive problem solving. Although being conceptually rather simple and easy to use there are nevertheless several things to be stated about this approach. First, a dynamic programming solution, though being recursively described, is not implemented recursively on a computer, but in an iterative manner, computation proceeding from solutions for smaller instances to solutions of larger instances. Only by this there is a chance to avoid the combinatorial explosion that would be the consequence of a direct top-down recursive implementation. What one has to deal with when applying dynamic programming is the question of how larger problem instances may be reduced to smaller problem instances, and of how solutions of these smaller problem instances may be assembled into a solution of the larger instance. This is most often rather canonical, but can sometimes be more sophisticated. There are a couple of further "tricks" a user should know, as for example building-in of constraints into the functions to be optimized, or avoidance of multiple evaluation of identical terms within function calls by computing and storing such terms in advance (before evaluation of recursive subterms starts) or at parallel to the main recursion.

Having seen dynamic programming as a design paradigm that leads to efficient and exact algorithms in a highly uniform manner ("cooking recipe"),

---

[1] for the reader...?

Chap. 4 contains much more sophisticated approaches. Here, we highlight the role that intelligent data structures and particularly well-chosen representation and visualization of problems play; a phenomenon that is widely known from, for example, mathematics, algorithm design, and artificial intelligence. We illustrate the enormous impact of proper representations on problem solution by four examples: the quite elegant data structure of PQ-trees that almost trivializes the CONSECUTIVE ONES problem occurring in DNA mapping; the powerful data structure of suffix trees with so many applications in bioinformatics that one is tempted to state "no suffix trees, no bioinformatics"; the astonishing way to make least common ancestor queries in arbitrarily large trees cost almost nothing by a clever single preprocessing of the considered tree; the usage of reality-desire diagrams as a visualization tool making the quite long, technical, and hard to read proof of the beautiful Hannenhalli/Pevzner formula for the reversal distance between signed permutations transparent and quite straight-forward.

Having thus presented more involved approaches in Chap. 4 that are by no means as uniform as the "cooking recipe" of dynamic programming from Chap. 3, this should remind us that we have to be aware that bioinformatics problems may be quite hard, as it is usually the case for most scientific fields dealing with algorithmic problems. Chapter 5 now turns to these hard to solve problems. The formal concept expressing hardness is the concept of *NP-completeness* that was independently introduced by Cook [22] and Karp [38] in 1971 who proved a first problem, satisfiability problem 3SAT for a restricted class of Boolean formulas, to be NP-hard. Showing that further problems are NP-hard is done by polynomially reducing a known NP-hard problem to the problems under consideration. Thus, in principle we obtain longer and longer reduction paths from initial problem 3SAT to further NP-hard problems. In the presentation of NP-hard problems in the literature there are often considerable gaps in reduction paths that a reader must accept. We completely avoid any such gaps by presenting for each NP-hard problem a complete reduction path from 3SAT to the problem under consideration. Chap. 5 also presents several NP-hardness proofs that are quite sophisticated, and thus seldom treated in textbooks but only cited from original papers.

Having learned now that we must be aware of an overwhelming number of NP-hard problems in bioinformatics that, as complexity theory tells us, cannot be efficiently solved by exact algorithms (unless P = NP), the question arises which sort of relaxed solution concepts could lead to efficient and practically useful algorithms also for the case of NP-complete problems. Here the idea of *approximation algorithms* is central. Such algorithms deliver solutions that deviate from an optimal one up to a certain constant degree, for example, they may guarantee that a computed solution is at most 10% more costly than a cheapest solution would be. Approximation algorithms are widely applied in bioinformatics to solve NP-hard problems, in particular in Holy Grail areas like sequencing, multiple alignment, and structure prediction. Chapter 6 presents a selection of approximation algorithms and makes clear that the idea of lower bounding the costs of a minimal solution, respectively of upper bounding the costs of a maximal solution, is at the heart of approximation algorithms.

Having so far presented mathematically well-founded approaches that lead to exact algorithms or at least to approximation algorithms with a provable guarantee on approximation factor, the wide field of *heuristic approaches* should not be completely omitted. It is often those surprising heuristic approaches which lead to the practically most useful solutions, seldom with an explanation why this is so and where the limits of such approaches are. In that sense, the theme of heuristics falls somehow outside the scope of a formally based book. This is reflected by the fact that we present in Chap. 7 only a small number of widely applied metaheuristics, and give only brief introductions (tutorials) to them. Having worked through these tutorials, the reader should be well prepared to read original papers or special textbooks on such heuristics - and she/he is encouraged to do so.

The book ends with a bibliography guiding the reader to a deeper study of matter.

Bibliographic Remarks

This book is clearly not intended to compete with the huge amount of excellent standard textbooks on bioinformatics, but to supplement these from the standpoint of formal problem analysis and algorithm design. Particular nice (to the personal taste of the author) introductions to the field of algorithmic bioinformatics are Clote & Backofen [20], Gusfield [31] (extensive and written in a formal and mathematically very concise manner), Pevzner [64] (a really fun to read book), Setubal & Meidanis [68], and Waterman [78] (the latter two books being truly classical bioinformatics textbooks). As almost all of the material treated in this textbook may also be found in any of these mentioned books, though sometimes these books do not present proofs or formal things up to the finest details, we almost completely omit to cite the books mentioned above (the alternative would be to steadily have citations from these books within the text). Concerning fields as complexity theory, approximation algorithms, Hidden Markov models, neural networks, or genetic algorithms we postpone recommendations for further readings to the end to the corresponding chapters. We cite original papers where the corresponding particular themes are first treated.

Errata and Links

There will be a steadily actualized errata website which can be found under www.inf.uos.de/theo/public. Further detected errors are always welcome under email address sper@informatik.uni-osnabrueck.de. A growing toolbox website is also available under www.inf.uos.de/theo/public. Here we provide for various demo tools that may serve the purpose to support understanding of algorithms presented in the book.

Acknowledgements

Jana Sperschneider (Albert-Ludwigs-Universität Freiburg, Germany and University of Western Australia, Australia) did a fourfold great job in the creation of this book: she did the laborious work to transform the manuscript into LATEXformat; she considerably improved overall presentation; she made "true English" from what I had written; she contributed all parts that cover RNA and pseudoknots. Lena Scheubert (Universität Osnabrück, Germany) contributed the chapter on genetic and ant algorithms. Thanks to many students to whom I had the opportunity to teach the themes of this book, and who improved presentations by steady criticism. Thanks also to all those students who took the opportunity to realize more or less extended implementation projects, some of which are presented on the bioinformatics toolbox website announced above. Last but not least thanks to the Springer team, in particular Hermann Engesser who supported the project from the beginning, Dorothea Glaunsinger and Frank Holzwarth who perfectly managed the production of the book.

Osnabrück, Germany,   *Volker Sperschneider*
March 2008

Contents

1

Core Bioinformatics Problems

Bioinformatics deals with the analysis, storage, manipulation, and interpretation of macromolecules such as DNA, RNA, or proteins. Rapidly improved techniques in biotechnology made a flood of analysed molecules accessible. Growing insight into the inner mechanisms of life allows for a meaningful interpretation of the information contained within these molecules and the explanation of the roles they play in the game of life. Conversely, the analysis of macromolecules further enhances our understanding of what is going on in cells on a molecular level. Progress in computer science makes it possible to deal with this giant flood of data in an efficient manner, with bioinformatics problems often triggering the development of new methods and algorithms.

We start with the description of a couple of core problems from bioinformatics, which will be thoroughly analyzed with respect to algorithmic solvability in later chapters: *DNA mapping and sequencing; string storage and string manipulation; pattern matching and string alignment; multiple alignment; gene finding; genome comparison; RNA structure prediction; protein structure prediction; regulatory network analysis.*

1.1 DNA Mapping and Sequencing

1.1.1 Size of Human DNA

Human genome consists of approximately 3.3 billion (i.e. 3.300.000.000) base pairs. These are organized into 23 chromosome pairs of size ranging up to a few 100 millions of base pairs. Let us make the naive assumption that we could read out bases of a human genome base by base with a rate of 1 second per base. Working 8 hours a day and 200 days a year it would take us $3.300.000.000/60 \times 60 \times 8 \times 200 = 572$ years to analyse a single human DNA molecule. It is well known that the Human Genome Project took a few years from its very beginning to the presentation of a successfully sequenced human DNA. Thus it appears that there must be a more efficient method than sequential base reading. Rapid sequencing of long DNA molecules is based on massive parallel processing of fragments of DNA along with the usage of computation and storage power of modern computers. Furthermore, the development of clever algorithms for the assembly of sequenced fragments to obtain (hopefully) their correct ordering on the original DNA molecule was essential. To understand this approach and the underlying algorithms we must take a short look into methods for the manipulation of DNA molecules used in biology laboratories.

1.1.2 Copying DNA: Polymerase Chain Reaction (PCR)

Working with DNA in laboratories requires lots of copies of the respective molecule. Fortunately, there is a clever process called *Polymerase Chain Reaction* which produces identical copies of a single DNA molecule in a cheap and fast manner. This reaction makes use of the built-in ability of DNA to reproduce itself. Under careful heating, double-stranded DNA splits into its two strands. With the presence of a sufficient amount of bases and the polymerase enzyme, each of the strands is rapidly complemented to a double-stranded copy of the original molecule. Repeating this process leads to an exponential increase in the number of identical copies, 2^n after the nth iteration.

1.1.3 Hybridization and Microarrays

Often one wants to know whether a certain string S occurs on either strand of some DNA molecule T. This can be answered by using again the basic property of DNA to complement a single strand, either as a whole as in PCR, or in parts in a process called *hybridization*. Hybridization of S to one of the strands of T can be detected as follows. Fix S onto a glass plate, then flood this glass plate with a solution of fluorescent (or radioactively marked) copies of T. After a while, wash out whatever has not hybridized to S and detect via fluorescence whether S has hybridized to a copy of T. Without major effort there may be some 10.000 test strings spotted onto a glass plate and simultaneously tested for being present on DNA string T. The testing works by making a photo of fluorescent spots which then is examined by automated methods. As an example, by working with lots of expressed mRNA strings one can obtain a complete snapshot of what is going on in a cell at a certain moment.

1.1.4 Cutting DNA into Fragments

Whenever only random fragments of DNA molecule copies are required, one may apply ultrasonic or press DNA through spray valves to achieve arbitrary cutting into fragments. There is another way of cutting DNA into fragments which works in a clearly controlled way, namely the use of restriction enzymes.

Each restriction enzyme cuts the DNA at every position where a certain short pattern, called restriction site, is present on either side of the DNA. In order to ensure that both strands are cut, it is required that wherever the restriction site occurs on one strand, it also has to occur on the other strand at the same position, but in reversed direction. Expressing this in string terminology, it means that restriction sites must be complementary palindromes, e.g. the string CAATTG. Assuming that the corresponding restriction enzyme cuts between the two bases A, the resulting DNA strings possess so-called *sticky ends*. In case that cutting is performed in the middle of the restriction sites, i.e. between the second base A and first base T, separation leaves so-called *blunt ends* (Fig. 1.1).

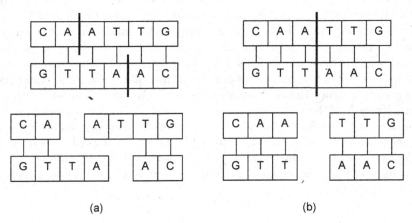

(a)   (b)

Fig. 1.1. DNA cleavage: **(a)** with sticky ends; **(b)** with blunt ends

1.1.5 Prefix Cutting

A special way of cutting DNA is to produce every initial segment ending with a fixed base, for example base A. This can be done by letting polymerase complement one strand of T from left to right in a solution of the four bases A, C, G, T and a suitable amount of slightly modified bases A that have the property to stop further extension[1]. Using a suitable proportion of modified bases A and sufficiently many copies of T one can guarantee with high confidence that the extension process will be stopped indeed at every position where the base A occurs.

1.1.6 Unique Markers

There are special DNA strings called *sequence tagged sites* (STS) which are known to occur exactly once on chromosomes. Sequence tagged sites can be

---

[1] In the following sections, take care to not confuse 'string' T with 'base' T.

retrieved from STS databases. Usually, they are hundreds of bases long and appear approximately every 100.000 bases on DNA. They are particularly helpful for getting a first rough orientation within a DNA molecule. In some sense they are comparable to unique landmarks in a foreign environment.

1.1.7 Measuring the Length of DNA Molecules

This is an easy task.

1.1.8 Sequencing Short DNA Molecules

DNA molecules up to a length of approximately 700 base pairs may be rapidly and automatically sequenced in the laboratory employing a method called *gel electrophoresis*. First, copies of the molecule under consideration are generated. Second, all of its prefixes ending with base A are constructed (as described before in Sect. 1.1.5). These prefixes are put onto a gel with an electric field applied to it. Charged prefixes are drawn through the gel, the shorter ones for a longer distance than the longer ones. This procedure is repeated three times, for each of the remaining bases C, G, and T. On the gel we obtain a pattern as in the example shown in Fig. 1.2. From this one can readily extract the sequence of bases in the DNA to be TTTGCCTAGGCTAT. Certainly, the approach is limited by the ability of gels to clearly separate strings up to a certain length. At the moment, this limit applies to strings of approximately 700 base pairs in length.

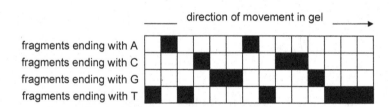

Fig. 1.2. Gel electrophoresis experiment.

1.1.9 Mapping Long DNA Molecules

Let us consider whole chromosomes, i.e. DNA molecules consisting of several hundreds of millions of base pairs. Usage of sequence tagged sites may help to give a first orientation of the structure of such molecules (see [37] for a map of human genome based on sequence tagged sites). For this purpose, we first produce long enough fragments of numerous copies of a chromosome T. Using hybridization, we then determine which subset of a given selection of sequence tagged sites appears on each fragment. Note that we do not obtain the correct ordering of sequence tagged sites on each fragment and we certainly have no knowledge about the correct ordering of fragments on the chromosome T. Nevertheless, in case that there are lots of fragment pairs sharing sufficiently many sequence tagged sites, we may use this redundancy to reconstruct both the correct ordering of sequence tagged sites on T and the correct ordering of fragments on T. We will discuss this approach in greater detail in the following chapters. At the moment we should state that this approach is expected to work only for rather long DNA molecules since it is required that a sufficient number of sequence tagged sites appears on the used fragments. For example, imagine we had exactly one sequence tagged site on every fragment. This obviously would give us no information at all on the true ordering of sequence tagged sites. We illustrate the idea of mapping DNA via sequence tagged sites through a very small example.

Assume we consider five sequence tagged sites $1, 2, 3, 4, 5$ and six fragments $F_1, F_2, \ldots, F_6$. Remember that neither the ordering of sequence tagged sites nor the ordering of fragments must correspond to their true ordering on DNA. Let $sts(F)$ be the set of sequence tagged sites appearing on fragment F. Assume that we observe the following scenario: $sts(F_1) = \{4\}, sts(F_2) = \{1, 5\}, sts(F_3) = \{1, 3\}, sts(F_4) = \{2, 4\}, sts(F_5) = \{2\}, sts(F_6) = \{1, 3, 5\}$. This can be visualized nicely with a matrix having a black entry in row i and column j in case that $sts(F_i)$ contains the number j (Fig. 1.3 (a)). This matrix clearly shows that we are not dealing with the true ordering of sequence tagged sites as there are rows containing gaps between black entries. The true ordering would not lead to rows with gaps between black entries. Rearranging sequence tagged sites into the ordering $2, 4, 5, 1, 3$ returns a matrix with consecutive black entries within each row (Fig. 1.3 (b)). Thus this scenario could possibly be the true ordering of sequence tagged sites. The question arises of how to find such an ordering, or even more, how to obtain all such orderings. Finally, arranging fragments according to their leftmost black entries leads to the matrix in Fig. 1.3 (c). Here, we have a plausible ordering of fragments, too.

1.1.10 Mapping by Single Digestion

The general principle behind mapping can be described as follows: produce sufficiently many copies of the DNA molecule T to be mapped and cut them into fragments. Afterwards, take a fingerprint from each fragment. For mapping using sequence tagged sites the fingerprint of a fragment F is the set of sequence tagged sites occurring on it. In case that fingerprints contain enough information about a fragment and there is sufficient redundancy within fingerprints we may try to infer the true ordering of fragments. One can utilize simpler fingerprints for the reconstruction of fragment ordering than sets of sequence tagged sites, e.g. the length of fragments. This is the basis for an approach called *single digest* mapping. In this approach, a single restriction enzyme E is used to cut the copies ("digest DNA"). The trick is to let enzyme E act on the copies of T only for a limited time such that it cannot cut off

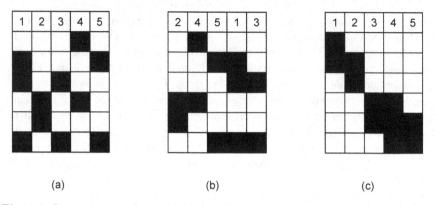

Fig. 1.3. Sequence tagged sites $1, 2, 3, 4, 5$ appearing on six fragments: **(a)** wrong ordering of STS's; **(b)** possible correct ordering of STS's; **(c)** possible correct ordering of fragments

every copy at every occurrence of its restriction site. By carefully controlling this experiment one can guarantee with high confidence that all segments of T are returned between any two occurrences of the restriction site of E. Clearly note that it is actually intended to obtain not only all segments between consecutive occurrences of the restriction site, but between any two occurrences. It is obvious why segments between consecutive restriction sites alone are no help for the reconstruction of fragment ordering: they do not involve any overlapping of fragments, thus do not contain any redundancy.

As an example, assume we have occurrences $O_1 < O_2 < O_3$ of the restriction site of E on a DNA molecule of length n. We obtain as fragments the segments between 1 and O_1, 1 and O_2, 1 and O_3, 1 and n (for a fragment that completely survived digestion), O_1 and O_2, O_1 and O_3, O_1 and n, O_2 and O_3, O_2 and n, O_3 and n. In contrast, a complete digestion would have led only to segments between 1 and O_1, O_1 and O_2, O_2 and O_3, O_3 and n. For each obtained segment its length is taken as fingerprint.

1.1.11 Mapping by Double Digestion

Now we let enzyme E act on the copies of DNA molecule T for a time sufficient to obtain all segments between consecutive occurrences of the restriction site of E and their lengths as fingerprints. As discussed above, this is no basis for the reconstruction of the true ordering of fragments. Things get better if we perform a second complete digestion using another enzyme F. This leads to a complete cutting of T at all occurrences of the restriction site of F. Again we determine all fragment lengths. Finally, in a third experiment we let both enzymes simultaneously cut the DNA at their corresponding restriction sites. As an example, consider the situation shown in Fig. 1.4 with cutting positions of enzymes E and F and fragment lengths indicated.

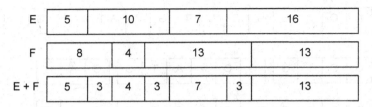

Fig. 1.4. Double digest experiment

Observe that the cutting positions in the third experiment result from combining cutting positions from the first two experiments. We look at the three lists of fragments lengths, sorted in ascending order: $L(E) = (5, 7, 10, 16)$, $L(F) = (4, 8, 13, 13)$, $L(E + F) = (3, 3, 3, 4, 5, 7, 13)$. We discover that several list rearrangements can be ruled out as being impossible. This raises hope that redundancy within the three lists may help to significantly reduce the number of admissible rearrangements. On the other hand, one gets the strong feeling that in order to find one or all admissible rearrangements, nothing better can be done than enumerating all permutations of $L(E)$ and $L(F)$ and see which of them leads to a permutation of $L(E + F)$. In case that $L(E)$ consists of n numbers we have to enumerate $n!$ many permutations, a number considerably greater than exponential 2^n. In the chapter on NP-hard problems we will indeed show that mapping by double digestion cannot be done faster (unless P = NP).

So far we have discussed three approaches for mapping long-sized DNA molecules with sufficiently long fragments (of at least a few hundred thousand base pairs). Now, we must finally think about methods to sequence DNA strings of mid-sized length. Note that for such sequences we can neither apply mapping by hybridization since sequence tagged sites may no longer be densely present on fragments. Nor can we use gel electrophoresis since we have not yet reached the required sequence length of at most 700 base pairs.

1.1.12 Mid-sized DNA Molecules: Shotgun Sequencing

Several copies of DNA molecule T are randomly cut into fragments of length up to 700 base pairs each. Fragments are automatically sequenced in the laboratory as described before (Sect. 1.1.8). Fragment overlaps can be expected in case we utilize sufficiently many copies and fragments that are not too short. These can then be used to reconstruct the string T with high confidence. As an example, Fig. 1.5 shows a couple of fragments that result from three copies of a DNA string.

A good strategy for the reconstruction of string T from fragments would surely be the following "greedy" strategy. We search for overlaps with maximum length between any two fragments and then melt these two fragments into a single string. Finally, this leads to a string S that contains every fragment string as a substring. Such a string is called a superstring for the frag-

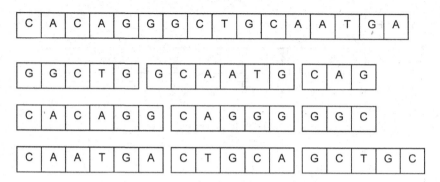

Fig. 1.5. Fragmentation of three copies of DNA into 9 fragments

ments. However, the string S obtained by this strategy must not necessarily correspond to the shortest possible superstring for the fragments. Here an algorithmic problem arises, namely to determine a shortest common superstring for the set of fragment strings. The reader may try to find out whether there is a common superstring for our example fragment set that is shorter than 16 bases. This constitutes the core problem in DNA sequencing. Indeed, one of the two competing attempts for sequencing a complete human genome was exclusively based on this approach (Celera Genomics, see [75]), whereas the other one also used DNA mapping as a first step (HUGO, see [21]).

From a complexity theory point of view, it will turn out that the computation of a shortest common superstring is a rather hard problem to solve. This will be shown in a later chapter. To demonstrate the dimension of this task in a real life situation, imagine that 10 copies of a human genome of size 3.3 billion base pairs are cut into fragments. Let us roughly estimate the amount of space required to store these fragments. For example, imagine DNA was printed in book form. Think of books of size 20 cm × 30 cm × 5 cm with 500 pages, 50 lines per page, and 80 characters per line. We require $33.000.000.000/500 \times 50 \times 80 = 16.500$ books. This leads to space requirements of about $50\,\mathrm{m}^3$, which corresponds to a box of size almost $4\,\mathrm{m}$ in each dimension filled with fragments. Enjoy this task of sorting fragments.

1.2 String Storage and Pattern Matching

1.2.1 Complexity of Pattern Matching in Bioinformatics

Working with strings in bioinformatics, in particular with DNA strings, we must be aware of the fact that huge numbers of strings have to be stored and processed, each of tremendous size. As a first example, consider the task of finding the maximum overlap (Fig. 1.6) between any two pairs of strings from a list consisting of a number of $f = 1000$ strings of length $n = 100.000$ each. This example drastically illustrates the complexity. There are f^2 pairs of strings to be treated. To compute the maximum overlap between any two strings we could naively test, for each k between 0 and n, by using k character comparisons whether there is a suffix-prefix match of length k in both strings. The maximum k with a suffix-prefix match of length k is the desired overlap length. For each string pair we thus have to perform $1 + 2 + \ldots + n = n(n+1)/2$ character comparisons. Taking the actual sizes from our example, we are confronted with $10.000.100.000.000.000/2 = 5 \times 10^{15}$ character comparisons.

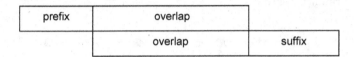

Fig. 1.6. Overlaps, prefixes, suffixes

1.2.2 First Look at Suffix Trees

We will later demonstrate that by using a very clever way to store strings the complexity of computing all overlaps between any two of the f fragments with length n can be reduced to the order of $fn + f^2$ steps (Chap. 5). This corresponds to $O(100.000.000 + 1.000.000) = O(10^8)$ for our example. The data structure that supports this task is called *suffix tree*. Given a string T of length n, its suffix tree organizes and stores all suffixes of T in a particularly efficient and intelligent manner. In some sense, it is comparable to an index of string T seen as a text. An index of text T contains links to all occurrences of important keywords, lexicographically and, to a certain degree, hierarchically ordered. Part of an index might contain the following entries:

> pattern
>> matching   211, 345, 346
>> recognition   33, 37, 88

This tells us that substrings 'pattern matching' and 'pattern recognition' occur on pages 211, 345, 346 and pages 33, 37, 88, respectively. Both share the prefix 'pattern', which leads to the tree-like organization. Of course, an index contains only a couple of important keywords occurring in text T. Contrary to this, a suffix tree contains all suffixes of text T and, for each such suffix, information about its starting position in text T. Searching for substrings is implicitly supported by a suffix tree, since arbitrary substrings of T can be found as prefixes of suffixes.

We take as an example the suffix tree of text $T = $ 'ananas' (Fig. 1.7). For ease of presentation trees are drawn from left (root) to right (leaves). One can identify every suffix of T as the path label of a path from the root down to some leaf. Don't worry about links occurring in the suffix tree at the moment. Their role will be explained in Chap. 4.

As the suffix tree of text T with length n contains n suffixes of T it is no shorter than T itself. There could even be the problem that representing suffixes with lengths ranging from 1 to n requires a tree of size $O(n^2)$. We will later see that this is not the case. Still, the problem will remain how to efficiently construct suffix trees. Despite of being a conceptually simple and easy to use data structure, naive approaches lead to complexities of $O(n^2)$ or even $O(n^3)$. Thus, the main focus in discussing suffix trees lies on efficient construction in time $O(n)$, whereas usage of suffix trees for lots of applications is rather straightforward.

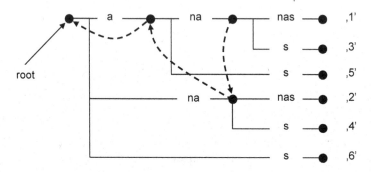

Fig. 1.7. Suffix tree for string 'ananas'

1.2.3 A Couple of Suffix Tree Applications

We mention a few applications of suffix trees together with a comparison of running times for a naive approach and a suffix tree approach. For each of the running times the reader may insert actual values for the parameters and observe how drastically running times improve:

- Find all occurrences of k patterns, each of length m, in a text of length n. A "very" naive approach has complexity $O(kmn)$, the well-known Knuth-Morris-Pratt algorithm that uses a preprocessing of the k patterns has $O(k(m + n))$ complexity, a suffix tree approach that uses a preprocessing of the text slightly improves complexity to $O(n + km)$.
- Find all overlaps between any two pairs of k strings, each of length n. The naive approach has complexity $O(k^2n^2)$, the suffix tree approach considerably improves this to $O(kn + k^2)$.
- For two strings of length n compute their longest common substring. The naive approach requires $O(n^3)$ time - give an explanation for this running time! Knuth conjectured that no linear time algorithm would be possible [44]. However, a suffix tree approach drastically reduces running time to $O(n)$ under proper (natural) measuring of complexity.

- Given the lower bound α and upper bound β determine all substrings with at least α and at most β occurrences in text T. Note that by properly setting α and β, this may return the interesting substrings occurring in T. We assume that strings with only a few occurrences are not characteristic, thus not interesting, for a text. Moreover, strings with overwhelmingly many occurrences are not interesting either since these are words like 'and' or 'to' which surely occur within any natural language text in abundance. Think of a naive approach and estimate complexity. It will surely not be linear, whereas a suffix tree approach can achieve $O(n)$ complexity.

Concluding this brief introduction of suffix trees we can state that they are a main driving force within bioinformatics tools: *"no suffix trees, no bioinformatics."*

1.3 String Alignment

There are at least three reasons why exact string matching applications as discussed in the section before are of secondary interest in bioinformatics. Consider a long sequenced string of DNA. First, we have to expect that measurements obtained in the laboratory are reliable only up to a certain degree. Second, even the process of typing data into computer storage must be seen as error prone up to a certain degree. Third, and of great importance is that nature itself is a source of non-exact matching between e.g. homologous strings in a sense that evolution may mutate bases, insert fresh bases, or delete existing bases. Thus it might be the case that we should not mainly look at the following three strings with respect to dissimilarity, but try to "softly match" them as best as we can.

```
T T A T G C A T A C C T C A T G G G T A C T
T A C G C G T A C C A T G T C A T T
T T A C G C G T A C T C A T G G T A C T T
```

This poses several questions. First question is what matching of strings might mean in case that not only exact coincidence of characters counts. This will lead us to the notion of a string alignment. For example, an alignment for the first and third string might look as follows.

```
T T A T G C A T A C – C – T C A T G G G T A C T
T T A C G C G T A C T C A T G G T A C – T – – T
```

Here we have introduced as few spacing symbols as possible in order to let as many characters as possible in both strings appear at the same position. Note that the introduction of a spacing symbol in one of the strings might be interpreted as a deletion of a formerly present character from this string, but also as an insertion of a fresh character into the other string.

Second question is of how to exactly measure the quality of an alignment. Here we must rely on a convention on how strongly to reward identities between characters and how to score mutations, insertions, and deletions. One

usual way for scoring is to look at valid aligned sequences obtained from biological insight or plausibility and count the following frequencies for each pair
(x, y) of characters or spacing symbol:

- p_{xy} = frequency of appearance of aligned pair (x, y)
- p_x = frequency of appearance of x

Now the relative frequency p_{xy} of an appearance of pair (x, y) is set into
relation to the probability of randomly aligning x with y measured by product
$p_x \, p_y$. Quotient $p_{xy}/p_x p_y$ expresses how strongly nature attempts to align x
with y, provided that x and y are sufficiently present. Note that a high value of
p_{xy} combined with an overwhelming presence of characters x and y (therefore
high values of p_x and p_y) may indicate muss less evolutionary pressure to align
x with y compared to a moderate value p_{xy} in combination with small values
p_x and p_y. Thus the quotient defined above is indeed the correct measure of
evolutionary preference for an alignment of x with y. For numerical reasons
one uses logarithms. Thus we obtain the following standard form of a scoring
function:

$$\sigma(x, y) = \log \frac{p_{xy}}{p_x \, p_y} \tag{1.1}$$

The closer quotient $p_{xy}/p_x p_y$ is to zero, the less probable it becomes to find
x aligned with y in regions with x and y being sufficiently present, and the
stronger pair (x, y) gets rewarded with a negative value. This is guidance
for alignment algorithms to avoid aligning x with y. Of course, a complete
alignment score consists of the sum of scores of its aligned character or spacing
symbol pairs. It can be easily shown that under mild and natural conditions
on a scoring function, they can always be written in the form above (1.1) -
at least after rescaling with a positive constant factor that does not influence
the quality of a scoring function. Of course, it is a delicate task to find enough
biologically reliable alignments as a basis for a plausible scoring function. As
an example, Fig. 1.8 shows the famous BLOSUM50 matrix for 20 amino acids
(each denoted by its 1-letter code). The spacing symbol does not occur within
this matrix as gaps are usually scored block-wise by a separate function, for
example by an affine function of block length.

Third question, finally, is how to efficiently compute alignments with maximum score. Simply enumerating alignments is not a good idea, since for strings
S and T of size $O(n)$ there exist $O(2^n)$ ways to introduce spacing symbols
in either one of the strings. Also, plain visual inspection of strings does not
help, not even for relatively short strings like the ones shown above: choose
some natural scoring function and try to get a feeling whether the first string
better aligns to the second or the third one. Therefore, we must develop efficient algorithms that compute optimal alignments. Chapter 3 on dynamic
programming will show such algorithms.

	A	R	N	D	C	Q	E	G	H	I	L	K	M	F	P	S	T	W	Y	V
A	5	-2	-1	-2	-1	-1	-1	0	-2	-1	-2	-1	-1	-3	-1	1	0	-3	-2	0
R	-2	7	-1	-2	-4	1	0	-3	0	-4	-3	3	-2	-3	-3	-1	-1	-3	-1	-3
N	-1	-1	7	2	-2	0	0	0	1	-3	-4	0	-2	-4	-2	1	0	-4	-2	-3
D	-2	-2	2	8	-4	0	2	-1	-1	-4	-4	-1	-4	-5	-1	0	-1	-5	-3	-4
C	-1	-4	-2	-4	13	-3	-3	-3	-3	-2	-2	-3	-2	-2	-4	-1	-1	-5	-3	-1
Q	-1	1	0	0	-3	7	2	-2	1	-3	-2	2	0	-4	-1	0	-1	-1	-1	-3
E	-1	0	0	2	-3	2	6	-3	0	-4	-3	1	-2	-3	-1	-1	-1	-3	-2	-3
G	0	-3	0	-1	-3	-2	-3	8	-2	-4	-4	-2	-3	-4	-2	0	-2	-3	-3	-4
H	-2	0	1	-1	-3	1	0	-2	10	-4	-3	0	-1	-1	-2	-1	-2	-3	2	-4
I	-1	-4	-3	-4	-2	-3	-4	-4	-4	5	2	-3	2	0	-3	-3	-1	-3	-1	4
L	-2	-3	-4	-4	-2	-2	-3	-4	-3	2	5	-3	3	1	-4	-3	-1	-2	-1	1
K	-1	3	0	-1	-3	2	1	-2	0	-3	-3	6	-2	-4	-1	0	-1	-3	-2	-3
M	-1	-2	-2	-4	-2	0	-2	-3	-1	2	3	-2	7	0	-3	-2	-1	-1	0	1
F	-3	-3	-4	-5	-2	-4	-3	-4	-1	0	1	-4	0	8	-4	-3	-2	1	4	-1
P	-1	-3	-2	-1	-4	-1	-1	-2	-2	-3	-4	-1	-3	-4	10	-1	-1	-4	-3	-3
S	1	-1	1	0	-1	0	-1	0	-1	-3	-3	0	-2	-3	-1	5	2	-4	-2	-2
T	0	-1	0	-1	-1	-1	-1	-2	-2	-1	-1	-1	-1	-2	-1	2	5	-3	-2	0
W	-3	-3	-4	-5	-5	-1	-3	-3	-3	-3	-2	-3	-1	1	-4	-4	-3	15	2	-3
Y	-2	-1	-2	-3	-3	-1	-2	-3	2	-1	-1	-2	0	4	-3	-2	-2	2	8	-1
V	0	-3	-3	-4	-1	-3	-3	-4	-4	4	1	-3	1	-1	-3	-2	0	-3	-1	5

Fig. 1.8. Blosum50 matrix

1.4 Multiple Alignment

Taking a protein family as an example, i.e. a set of related strings of amino acids, a multiple alignment (see [16] for general information on multiple alignment in the context of bioinformatics) might look as in Fig. 1.9. Note that the primary goal is not to obtain maximum similarity between pairs of strings, but to identify subsections of the proteins that are strongly conserved within all strings. Usually, such highly conserved subsections are those that define the biological function of the proteins within the family. Conversely, the remaining sections are those that have experienced variation during evolution without leading to an extinction of the species since they are not essential for biological function. For example, observe a strongly conserved 'CG' pattern at positions 7 and 8 of the alignment, as well as a strongly conserved 'PNNLCCS' pattern at positions 18 to 24. With a fixed scoring function, a multiple alignment may be scored by summing up all scores of pairwise combinations of aligned strings. This way of scoring a multiple alignment is called *Sum-of-Pairs scoring (SP-score)*. It is not as obvious as for the case of pairwise alignment that this is a reasonable way for scoring a multiple alignment. There are arguments indicating that SP-scoring is, in some sense, doubtful.

From a complexity theory point of view we will later show that finding a multiple alignment with maximum SP-score is NP-hard, thus there is great evidence that it is not solvable by an efficient algorithm. An alternative view to multiple alignments is to say that a multiple alignment defines some sort of *consensus string*, i.e. a string most similar to all aligned strings. Defining a consensus string may be done by different approaches, which nevertheless will

```
VAIAEQCGRQAGGKLC-PNNLCCSQWGWCGSTDEYCSPDHNCQSN-CK-
TAHAQRCGEQGSNMEC-PNNLCCSQYGYCGMGGDYCGKG--CQNGACYT
ATNAQTCGKQNDGMIC-PHNLCCSQFGYCGLGRDYCGTG--CQSGACCS
LVSAQRCGSQGGGGTC-PALWCCSIWGWCGDSEPYCGRT--CENK-CWS
TAQAQRCGEQGSNMEC-PNNLCCSQYGYCGMGGDYCGKG--CQNGACWT
TAQAQRCGEQGSNMEC-PNNLCCSQYGYCGMGGDYCGKG--CQNGACWT
----QRCGEQGSGMEC-PNNLCCSQYGYCGMGGDYCGKG--CQNGACWT
TVKSQNCG-------CAP-NLCCSQFGYCGSTDAYCGTG--CRSGPCRS
SAE--QCGRQAGDALC-PGGLCCSSYGWCGTTVDYCGIG--CQSQ-CDG
PAAAQNCG-------CQP-NFCCSKFGYCGTTDAYCGDG--CQSGPCRS
PAAAQNCG-------CQP-NVCCSKFGYCGTTDEYCGDG--CQSGPCRS
SAE--QCGQQAGDALC-PGGLCCSSYGWCGTTADYCGDG--CQSQ-CDG
SAE--QCGRQAGDALC-PGGLCCSFYGWCGTTVDYCGDG--CQSQ-CDG
----EQCGRQAGGKLC-PNNLCCSQYGWCGSSDDYCSPSKNCQSN-CK-
```

Fig. 1.9. Example multiple alignment exhibiting several strongly conserved subsections

be later shown to be equivalent. A first attempt is to look at an alignment column by column and take for each column the character (or spacing symbol) that has maximum summed score with all characters in the considered column. In the example above, such a consensus line would probably be the string

TAAAQCGEQGAGGLC–PNNLCCSQYGYCTTDDCG–CQSACWT.

By deleting spacing symbols we obtain a string with maximum summed alignment score with all strings occurring in the lines of the alignment. Such a most similar string is known in algorithm theory as *Steiner string*. Conversely, having a Steiner string S for a set of strings $S_1, \ldots, S_k$ we will later show how to construct, guided by S, a multiple alignment $T_1, \ldots, T_k$ for $S_1, \ldots, S_k$ whose consensus line coincides with S after deletion of all its spacing symbols. Thus we observe that constructing a multiple alignment with maximum score consensus line for strings $S_1, \ldots, S_k$ is equivalent to finding a common ancestor for strings $S_1, \ldots, S_k$ with maximum similarity to all of the strings.

multiple alignment ↔ *star-like phylogenetic tree with common ancestor*

The tree structure associated with a Steiner string (though only a trivial star-like tree is shown in Fig. 1.10) suggests the following generalization. Assume we had knowledge about various ancestors that occurred during evolution from a common root towards leaves $S_1, \ldots, S_k$. As an example consider the tree structure shown in Fig. 1.11.

Now it makes sense to ask for an assignment of strings to the four ancestor nodes a, b, c, d in the diagram such that the sum of all optimum alignment scores between any two strings connected by a link is maximized. This problem is called *phylogenetic alignment problem* in the bioinformatics literature; in algorithm theory it is known as *Steiner tree problem*. All of the problems treated here, i.e. multiple alignment with optimum consensus line or Steiner string and Steiner tree problems, do not allow efficient algorithms (unless P = NP). Two relaxations behave better. First, there are efficient algorithms that

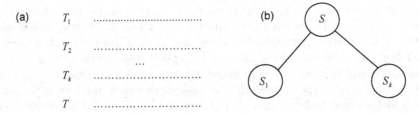

Fig. 1.10. (a) Multiple alignment with consensus line T; **(b)** Steiner string S and star-like tree

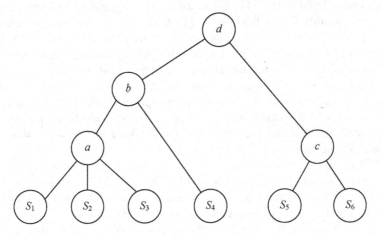

Fig. 1.11. Steiner tree/Phylogenetic alignment

guarantee to deliver solutions which are at least half as good as the optimal solution. Such methods are called 2-approximation algorithms [3]. Second, by relaxing the notion of a multiple alignment to the notion of a profile of an alignment, i.e. a distribution of frequencies for all characters within each column of an alignment, there exist efficient algorithms that optimally align a string to already obtained profiles (for example at node b), or an already obtained profile to another one (that happens at node d of Fig. 1.11).

So far we have primarily discussed themes that look at DNA strings, that is, *the book of life* as a sequence of characters. Books, of course, are structured into chapters, sections, sentences, and words. Guided by this analogy we should ask what might be the words in the book of life. This leads us to the discussion of algorithmic problems in the context of genes, i.e. protein-coding subsections of DNA.

1.5 Gene Detection

Having sequenced DNA strings the next important task is to identify functional units, primarily genes. Due to lack of knowledge of how to precisely recognize exons within a gene, or promoter regions of genes, gene detecting systems do not work by fixed deterministic rules, but are usually based on methods of statistics or adaptive systems such as neural networks. Such systems may automatically learn (more or less) good rules characterizing exons from a sufficient number of known exons. We will discuss such methods in the chapter on adaptive approaches. Here we concentrate on a different approach, which utilizes that biology knows about several indications for a substring to be part of an exon within a gene. For example, absence of a STOP codon within at least one of the three reading frames is required for an exon. Introns between exons exhibit the GT-AG motif at the beginning and end. Exons show a considerably higher frequency of CpG pairs than non-exon regions do. Using such indications (and further ones) we may predict exons with high sensitivity (all true exons are predicted to be exons), however with low specificity (we must tolerate exon prediction for strings that are not true exons). Nevertheless, such a sensitive list of exon candidates may be used to computationally assemble a selection of non-overlapping candidate exons with best alignment score to a protein string T that is hypothesized to be the one expressed by the gene under consideration. By letting T range over a database of protein strings one may find the one that can best be assembled by the candidate exons. This situation is denoted in Fig. 1.12. A selection of non-overlapping exon candidates is indicated by black shaded regions. As with all computational tasks discussed so far, a naive approach is not feasible: for a candidate list consisting of b strings one has to search through 2^b many selections of candidates to find the one whose concatenation optimally aligns to string T. A dynamic programming approach will perform considerably better, in particular run in time linear in the number b of candidate exons.

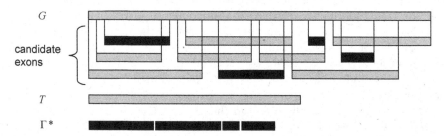

Fig. 1.12. A selection of non-overlapping candidate exons intended to best assemble target string T

1.6 Gene Comparison

Having identified genes in DNA strings of various organisms a comparison of DNA strings on basis of the ordering of genes (assuming that two organisms are equipped with the same genes) instead of single bases may be appropriate. This is interesting, for example, in case organisms exhibit very few point mutations and experience no loss or gain of genes at all. Variations of gene ordering may be caused by *reorganisation events* in the recombination process. The simplest such event, *directed reversal*, happens whenever a substring of DNA is build into DNA in a 180° rotated order (Fig. 1.13). Note that genes change their strand, whereas taking the mirror image of the strand does not make sense since it violates the chemical orientation of strands.

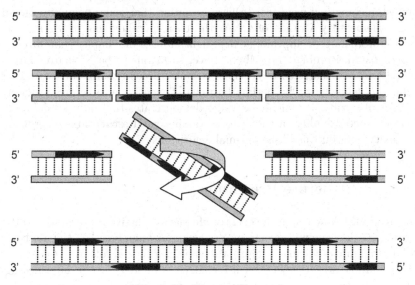

Fig. 1.13. Directed Reversal

Besides directed rearrangements within single chromosomes there are further known reorganisation events, e.g. translocations (treated in [40]) of genes or fusion and fission between two chromosomes. Assuming a constant (and known) rate of occurrence for such events, the minimum number of events required to transform the genome of one species into the genome of another one, the so-called *reversal distance*, would allow an estimation of how distant in evolution two species are. At first sight, the determination of reversal distance between two species with n genes each is not an easy task. Of course, there is a trivial upper bound on reversal distance, namely $2n$. This can be illustrated as follows: the first gene of the target genome can be put from its position within the initial genome to its correct place with at most two reversals: the first one bringing it to the front of the genome, the second one eventually moving it from one strand to the complementary strand. Thus, $2n$ reversals are sufficient to transform an initial order of n genes into a target order. Theoretically, we might try all reversal sequences of length at most $2n$ in order to find the shortest one that transforms initial order into target order.

Since a single reversal is characterized by two gene positions (position of the leftmost and rightmost gene within the reversed area), for every reversal step we have to try $O(n^2)$ many choices. In summary, $O((n^2)^{2n}) = O(n^{4n})$ reversal sequences must be explored. Taking as a realistic example mitochondrial DNA with 36 genes, e.g. in case of man and mouse, a search space of size $O(36^{144})$ has to be explored. This is approximately $O(10^{220})$, a number that is considerably greater than the estimated number of 10^{80} atoms in the universe. Astonishingly, Hannenhalli & Pevzner found an efficient algorithm for computing the reversal distance that works in time $O(n^2)$ (see [33] and [32]). Even more astonishingly, if we discard the information that genes belong to one of the complementary strands of DNA, the corresponding *undirected reversal distance* problem that appears, at first sight, a little bit simpler than *directed reversal distance* problem described above, turns out to be NP-hard, thus is probably not efficiently solvable. This is quite a nice example of two rather similar problems at the borderline between efficiently solvable and probably not efficiently solvable which drastically demonstrates that only a solid theoretical competence offers a chance of obtaining valid estimations of problem complexity and, on this basis, optimal algorithms.

1.7 RNA Structure Prediction

Contrary to DNA, which in the cycle of life plays a passive role as data storage medium, RNA also exhibits active roles as an enzyme. For this, the ability of RNA to fold into particular conformations plays a crucial role. As an example, a cloverleaf-like structure is typical for tRNA, which acts as an intermediary between codons of mRNA and corresponding amino acids according to the genetic code. Folding of single-stranded RNA is governed by complementary bases A-U and C-G, which form stable base pairs via hydrogen bonding. These two base combinations are often referred to as *Watson-Crick pairs*[2]. Base pairs are the foundation for secondary structure formation in single-stranded RNA. Domains of base pairs form characteristic structure elements, which are called *stems* or *helical regions*, *hairpin loops*, *bulges*, *internal loops*, and *multiloops*, all shown in an artificial example in Fig. 1.14.

In our example, structure elements do not interleave, thus exist either side by side or nested. Such "context-free" folds allow simple prediction algorithms based on a local optimization strategy. They all use dynamic programming and either attempt to maximize the number of base pairs or minimize the so-called free energy of a fold. Interleaving structure elements are known under the name of *pseudoknots* (Fig. 1.15). Pseudoknots play an important role in nature but require a much more sophisticated attempt to reliably predict them.

---

[2] Additional base pairs are often considered in the literature, e.g. the so-called *Wobble pair* G-U.

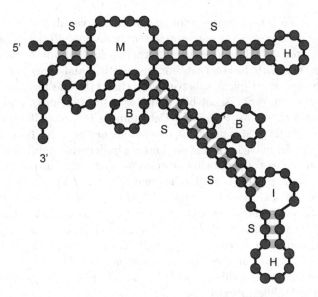

Fig. 1.14. Typical structures in a pseudoknot-free fold. Structure elements shown are stems (S), hairpin loops (H), bulges (B), internal loops (I), and multiloops (M) *(covalent bonds are indicated by solid lines, hydrogen bonds by shaded areas)*

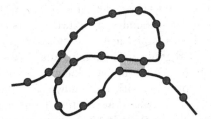

Fig. 1.15. Pseudoknot structure

1.8 Protein Structure Prediction

1.8.1 Holy Grail Protein Structure Prediction

Besides DNA as information storage medium and RNA with its various roles in transcription and translation, as well as enzymatic roles, there exists a third important group of macromolecules, namely proteins. As with RNA, functionality of a protein is mainly defined by its 3D fold. However, other than with RNA, protein folding is driven by a greater diversity of forces than mere hydrogen bonding. Usually much longer than RNA molecules, a greater variety and complexity of folds is observed for proteins. Predicting folds is thus much more difficult for proteins than for RNA and requires more advanced methods (see [23]). Due to its importance as well as its difficulties, obtaining good protein structure prediction methods thus counts as one of the Holy Grails of bioinformatics. To get an overview of the enormous variability of protein 3D structures the reader may try out protein structure presentation tools such as 'Protein Explorer' or others (use the internet to find an appropriate protein explorer together with instructions for installation and usage). Looking at proteins one observes typical substructures that take the form of helix-like folds (*alpha-helix regions*), sheet-like folds (*beta-sheet regions*), and unfolded regions between them (*loop regions*). Determining the exact 3D structure of a protein (its *tertiary structure*) in the laboratory may be done by X-ray crystallography. Drawbacks of this method are high costs, mainly due to time-consuming procedures of producing proteins in crystalline form and difficulties to crystallize some proteins at all. Thus one asks for computational methods for the prediction of tertiary structures. There are a lot of methods that differ in terms of the granularity of what they attempt to predict. Modelling all known forces that influence the process of protein folding one might try to simulate the complete folding of a protein with a computer starting with the sequence of amino acids (its *primary structure*). Due to the enormous complexity of the process this succeeded, up to date, only for very short time intervals of the folding process.

1.8.2 Secondary Structure

Another attempt on the way towards determining the tertiary structure of a protein is to first predict its secondary structure, i.e. its string of labels indicating to which of the classes 'alpha-helix', 'beta-sheet', or 'loop' every amino acid belongs. This is a good application area for adaptive systems such as neural networks since there is limited analytic knowledge on how folding works, whereas there are lots of proteins with well-known secondary structure. There are indeed neural network systems that attempt to predict the correct label of an amino acid in the middle of a window of limited width $2k + 1$ of known amino acids. This may be done with a rather simple feedforward architecture that uses $2k + 1$ groups of 20 input neurons to encode $2k + 1$ amino acids (each of the 20 amino acids is encoded by 20 neurons in a unary manner), a hidden layer of suitably many neurons, and three output neurons that encode in unary manner the three classes 'alpha-helix', 'beta-sheet' and 'loop'. Training it with sufficiently many input windows for which the correct class of its middle amino acid is known usually gives goods results for unknown windows. This approach requires a sufficient number of training instances whose correct labelling is already known (for example from X-ray crystallography), thus it may be called a *comparative approach*.

1.8.3 Tertiary Structure Prediction with Contact Maps

Behind comparative approaches lies the assumption that sequences with high similarity as primary sequences also exhibit high similarity on higher levels of description, up to the stage of biological function. An interesting comparative approach more closely related to tertiary structure is to predict the so-called *contact map* of an amino acid sequence. Here, for any two of its amino acids one wants to predict whether they are close in the native fold of the protein. Thus, for an amino acid sequence of length n, its contact map is an $n \times n$ binary matrix. The entries of the contact map of a protein up to a certain confidence allow a rather good reconstruction of its 3D structure. Also, prediction of contact maps with neural networks is easier and much better suited than determination of the exact geometry of a fold is. Again, neural networks are applied to predict each entry of this matrix. As input information one can make good usage of secondary structure information, as well as of correlations in mutational events using databases of homologous proteins at positions where contact/non-contact has to be predicted. Details are described in a later chapter.

1.8.4 HP-Model Approach

A related method, though not being a comparative approach but a de novo prediction of contacts, is the so-called *HP-model* approach. Here, amino acids are divided into hydrophobic and polar (hydrophilic) ones. Then, a fold of the protein into a square or cubic grid is searched that exhibits many non-consecutive pairs of hydrophobic amino acids in direct contact. Behind this is the imagination that in an aqueous environment hydrophobic amino acids tend to agglomerate in order to be as far away from water molecules, which is the main driving force in protein folding. For the sequence HHPHPHHPHPHHPHHHPPH, as an example, the left hand fold has 4 non-consecutive H-H pairs, whereas the right hand one has 6 non-consecutive H-H-pairs (Fig. 1.16). The question arises which is the maximum possible number of H-H-pairs in a fold. It is known that this problem is NP-hard thus probably not efficiently solvable. A well-known efficient approximation algorithm exists which determines the maximum number of H-H-pairs (as well as a fold realizing this number) up to a constant factor [35].

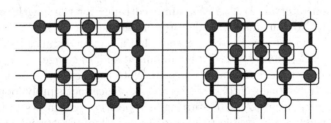

Fig. 1.16. Two HP-model folds with hydrophobic nodes in black and polar nodes in white

1.8.5 Protein Threading

A completely different approach is followed in *protein threading*. Here, one tries to map the amino acids of a novel protein to the helix, sheet, and loop structures of a protein with known 3D structure such that it results in maximum similarity within helix and sheet regions, also called *core segments*. The idea behind this approach is that core segments, which determine the biological function of homologous proteins, exhibit much less mutational changes than loop segments. Thus, whenever a novel protein can be nicely threaded through the core segments of a known protein there are good reasons to assume that both proteins exhibit similar 3D structures. Depending on how similarity of core segments is measured, a solvable or an NP-hard problem arises. Let $S = S(1)S(2)\ldots S(n)$ be the primary structure of a novel protein. Let a reference protein $T = T(1)T(2)\ldots T(k)$ be given with known tertiary structure consisting of a sequence of core segments (alpha or beta) $C_1, \ldots, C_m$ of fixed lengths $c_1, \ldots, c_m$ that are separated by loop segments $L_0, L_1, \ldots, L_{m-1}, L_m$ of lengths $l_0, l_1, \ldots, l_{m-1}, l_m$ (Fig. 1.17). Thus, reference protein T is segmented into $T = L_0 C_1 L_1 C_2 \ldots C_m L_m$.

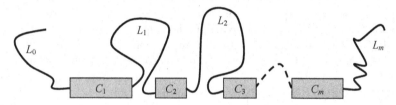

Fig. 1.17. Loop-core structure of reference protein T

We do not assume any knowledge about the exact nature of core segments. The only assumption made is that lengths of core segments are conserved in evolution whereas mutational events are allowed. Concerning loop segments we assume that their lengths are variable to a certain degree, which is expressed by an interval $[\lambda_i \ldots \Lambda_i]$ as admitted range for the length of the ith loop segment. Now the task is to identify core segments $C^1, \ldots, C^m$ of lengths $c_1, \ldots, c_m$ within the novel protein S such that a certain distance measure between $C_1, \ldots, C_m$ and $C^1, \ldots, C^m$ is minimized and, of course, constraints on the lengths of loop segments are respected. Note that, even if it is demanded that C^i exactly matches C_i (for $i = 1, \ldots, m$), one cannot simply look for the left-most occurrence of C_1 in T starting at a position within $[\lambda_1 \ldots \Lambda_1]$ since this might make it impossible to realize the next loop constraint $[\lambda_2 \ldots \Lambda_2]$. Thus, even for this special (and rather simple) case of exact matching between corresponding core segments we must look for more clever algorithms, not to speak about more elaborated ways of measuring similarity between core segments.

1.9 Molecular Networks

Gene regulation is a difficult theme, both with respect to single gene regulation and of course even more so for ensembles of interacting genes (pathways). Finding such regulations may be supported by the availability of huge amounts of microarray data. Fig. 1.18 shows a really small scenario of 24 measurements, each with the expression of 40 genes measured. Can you recognize a partition of the rows into two groups with similar expression behaviour inside each group with respect to a selection of genes? It is probably hard to detect any structure in the wild mixture of black and white entries. One might hope that algorithmic approaches might help to shed some light on such collections of data.

Fig. 1.18. Microarray measurements; 24 measurements (rows), 40 genes (columns)

The matrix in Fig. 1.18 has two submatrices (in Fig. 1.19 made visible after a permutation of rows followed by a permutation of columns), one consisting of 9 measurements and 16 genes expressed in all of them, the other one consisting of 11 measurements and 6 genes expressed in all of them. Finding such submatrices completely filled with black entries gives rise to saying, for example, that there are two clearly separated groups of measurements with associated genes in each group that explain what is different within both groups. Finding such homogeneous submatrices is referred to as *bi-clustering*: measurements are grouped into clusters, and at the same time genes are clustered into groups to obtain a rule explaining the division of measurements into clusters.

Fig. 1.19. An example of bi-clustering

1.10 Bibliographic Remarks

For getting an overview on main bioinformatics themes we especially recommend Setubal & Meidanis [68] and Pevzner [64].

2

Turning to Algorithmic Problems

2.1 Presentation of Algorithmic Problems

Having introduced a couple of core bioinformatics problems in the first chapter we now transform these problems into algorithmic problems using terminology that is well-known to a computer scientist. Usually, algorithmic bioinformatics problems are *optimization problems*. The following components characterize any optimization problem:

- An alphabet Σ of characters.
- A set I of admissible character strings that may occur as input instances.
- A solution relation S defining what is to be found, given admissible input x.
- An objective function c evaluating quality of solutions.
- A mode, *min* or *max*.

The role of these components and required properties of components are explained next. For being executable on a computer, all objects occurring within an algorithmic problem must be encoded in computer readable format. This is achieved by encoding data by strings over a suitable alphabet Σ. Usually only a subset of strings represents admissible data. For example, $\neg(x \lor y)$ encodes an admissible Boolean formula, whereas $\lor \neg x \neg y$ is not admissible. This is expressed by defining an instance set I of admissible input strings x. Of course, it should be possible to efficiently test whether strings x are members of I. Given admissible string x one is interested in strings y serving as solutions for x. What exactly solutions are is represented by a binary relation $S(x, y)$ on pairs of strings. As with instances, it should be possible to efficiently test whether strings are solutions for admissible strings. Furthermore, any solution y for x should be bounded in length by a polynomial in the length of x. This makes sense since otherwise size of solutions would already prevent efficient storing or printing of solutions. What a solution is for a concrete example varies considerably: a solution for an equation or system of equations; a satisfying truth value assignment for the variables of a Boolean formula; a path in a graph; and many more. Solutions are scored via an objective function $c(x, y)$ assigning a numerical value to any pair (x, y) consisting of admissible string x and solution y for x. Of course, $c(x, y)$ should be efficiently computable. Finally, an optimization problem has a mode, either *max* or *min*. For a minimization problem and admissible instance x define:

$$c_{\mathrm{opt}}(x) = \begin{cases} \text{`no solution'} & \text{if no solution exists for } x \\ \min_{\text{all solutions } y \text{ for } x} c(x, y) & \text{otherwise.} \end{cases} \qquad (2.1)$$

For a maximization problem, *min* is replaced by *max*. Minimization problems are always introduced in the following schematic format:

> **NAME**
> Given admissible x, find a solution y for x with minimum value $c(x, y) = c_{\mathrm{opt}}(x)$.

Sometimes we concentrate on the following simplified problem version that only computes the optimal value of a solution, but does not return an optimal solution.

> **NAME**
> Given admissible x, compute value $c_{\mathrm{opt}}(x)$.

Usually, an optimal solution y for x can be extracted from the computation of minimum value $c_{\mathrm{opt}}(x)$ with limited extra efforts. Sometimes there is no objective function to be optimized, only the task to simply find a solution provided one exists. This can be seen as a special form of optimization problem with trivial objective function $c(x, y) = 1$, for every solution y for x.

2.2 DNA Mapping

2.2.1 Mapping by Hybridization

Usage of sequence tagged sites for hybridization with overlapping fragments of the DNA molecule to be mapped was described in Sect. 1.1. The resulting algorithmic problem is the problem of rearranging columns of a binary matrix with the objective to make all bits '1' appear in consecutive order within every row.

> **CONSECUTIVE ONES**
> Given a binary matrix with n rows and m columns, find all possible rearrangements of columns that lead to a matrix with all bits '1' occurring in consecutive order within each row.

The algorithmic solution that will be presented in Chap. 4 uses the concept of *PQ-trees* introduced by Booth and Lueker (see [13]) . To motivate this idea with an example, assume that rows 1 and 2 of a binary matrix contain common bits '1' in columns 1, 2, 3, row 1 contains additional bits '1' in columns 4, 5, 6, and row 2 contains additional bits '1' in columns 7, 8. Let 9, 10, 11 be the remaining columns which thus contain common bits '0' in rows 1, 2. To put bits '1' in columns 1, 2, 3, 4, 5, 6 of row 1, as well as bits '1' in columns 1, 2, 3, 7, 8 of row 2 in consecutive order, we are forced to rearrange columns as follows: take an arbitrary permutation of columns 4, 5, 6, followed by an arbitrary permutation of columns 1, 2, 3, followed by an arbitrary permutation of columns 7, 8. Alternatively, an arbitrary permutation of columns 7, 8, followed by an arbitrary permutation of columns 1, 2, 3, followed by an arbitrary permutation of columns 4, 5, 6 may be performed. In either variant, columns 9, 10, 11 may be arbitrarily arranged besides one of the described arrangements of 1, 2, 3, 4, 5, 6, 7, 8. The reader surely will have noticed that this description is somehow difficult to read. The following concept of a tree with two sorts of nodes, P-nodes and Q-nodes, simplifies descriptions considerably. A *P-node* is used whenever we want to express that an arbitrary permutation of its children is admitted, whereas a *Q-node* is used whenever we want to express that only the indicated fixed or the completely reversed order of its children is admitted. P-nodes are drawn as circles, whereas Q-nodes are drawn as rectangles. Using this notion, the arrangements described above can be visualized as shown in Fig. 2.1.

An algorithm described in Chap. 4 will successively integrate the requirements of consecutiveness, row by row, as long as this is possible. It starts with the PQ-tree consisting of a single P-node as its root and n leaves that does not constrain order of columns. Ideally, the algorithm ends with a maximally fixed PQ-tree consisting of a single Q-node as root and a permutation of column indices at its leaves. Note that a complete reversal of ordering is always possible in every PQ-tree. Thus, data may determine ordering of columns only up to such a complete reversal. Whenever the algorithm returns a more complex PQ-tree this indicates that the data was not informative enough to uniquely fix (up to a complete reversal) ordering of columns.

2.2.2 Mapping by Single Digestion

Given a string of length n and ascending positions (including start and end positions 1 and n) $1 = x_1 < x_2 < \ldots < x_m = n$, define its single digest list as the sorted list of all differences $x_p - x_q$, with indices p and q ranging from 1 to m and $p > q$. Note that for a list of m positions, its single digest list consists of a number of $m(m-1)/2$ entries (duplications allowed). The single digest problem is just the computation of the inverse of the single digest list function.

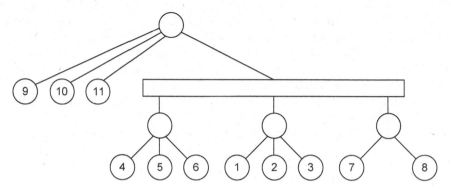

Fig. 2.1. Example PQ-tree

> **SINGLE DIGEST**
> Given a sorted list L of m lengths (duplications allowed) with maximum length n and minimum length 1, find all possible ascending lists P of positions ranging from 1 to n, (always including 1 and n) with single digest list of P identical to L.

We illustrate an algorithm solving the problem on the basis of a complete backtracking search through all possible position lists P with a very simple example. After this, we summarize what is known at the moment about the complexity theory status of the problem. As an example, consider the list of lengths $L = [2, 2, 3, 7, 8, 9, 10, 11, 12, 17, 18, 19, 21, 26, 29]$ consisting of $1/2 \times 5 \times 6 = 15$ entries. A suitable list of positions $P = [0, x_1, x_2, x_3, x_4, 29]$ has to be found such that its single digest list coincides with L, provided such a list exists.

We start with position list $P = [0, 29]$ that realizes the difference length 29. Lengths $2, 2, 3, 7, 8, 9, 10, 11, 12, 17, 18, 19, 21, 26$ remain to be produced by setting further cut positions. From now on, we always concentrate on the greatest number z amongst the remaining lengths which have not yet been realized as a difference between positions. Obviously, z can only be realized either relative to the left border as difference $z = p - 0$ or relative to the right border as difference $z = 29 - p$ with a further cut position p, since any other way would result in a number greater than z. Thus we have to search the binary tree of all options in order to set the next cutting position either relative to the left or the right border. In Fig. 2.2, branches L represent realization $z = p - 0$ relative to the left border, whereas branches R represent realization $z = 29 - p$ relative to the right border. This tree is traversed in a depth-first manner for as long as the next cut position can be set consistently with the actual list of remaining lengths. If the setting of the next position is inconsistent with the actual list of remaining lengths, backtracking takes decisions back and proceeds tree traversal at the next available unexplored branch of the tree. As we will see for the chosen example, lots of subtrees may be pruned, i.e. they must not be visited in the traversal. Such pruned subtrees are indicated in Fig. 2.2 by enclosing them in shaded rectangles.

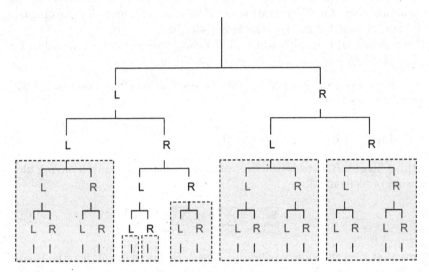

Fig. 2.2. Tree of all settings of cut positions showing pruned sub-trees that represent inconsistent settings

We show which positions are tried out at any stage of the traversal, which lengths remain to be realized at any step, and which subtrees are pruned due to leading to inconsistent lengths.

branch	positions	remaining lengths to be realized
root	0,29	2,2,3,7,8,9,10,11,12,17,18,19,21,26
L	0,**26**,29	2,2,7,8,9,10,11,12,17,18,19,21
LL	0,**21**,26,29	inconsistent due to $26 - 21 = 5$
LR	0,**8**,26,29	2,2,7,9,10,11,12,17,19
LRL	0,8,**19**,26,29	2,2,9,12,17
LRLL	0,8,**17**,19,26,29	inconsistent due to $17 - 8 = 9$ and $26 - 17 = 9$
LRLR	0,8,**12**,19,26,29	inconsistent due to $12 - 8 = 4$
LRR	0,8,**10**,26,29	inconsistent due to $26 - 10 = 16$
R	0,**3**,29	2,2,7,8,9,10,11,12,17,18,19,21
RL	0,3,**21**,29	inconsistent due to $3 - 0 = 3$
RR	0,3,**8**,29	inconsistent due to $8 - 3 = 5$

We observe that list L cannot be represented as the single digest list of a position list P. To discover this, considerable parts of the tree had not to be visited. This massive pruning effect can also be observed in implementations. It sheds some light on the first one of the following remarks on the complexity theory status of the problem.

- The backtracking procedure described above usually visits only rather limited parts of the complete search tree to either find a position list P with single digest list L, or to find out that such a realization is not possible. Informally stated, average running time is low. There are more precise estimations found in the literature on what is to be expected, on average, as execution time for the backtracking procedure.
- No polynomial time algorithm solving the problem has been found so far.
- The problem has not been shown to be NP-hard so far.

Thus, a lot of work is left concerning the determination of the exact complexity status of the problem.

2.2.3 Mapping by Double Digestion

Let three lists A, B, C (duplications allowed) of lengths be given having identical sum of elements, i.e.

$$\sum_{a \in A} a = \sum_{b \in B} b = \sum_{c \in C} c. \tag{2.2}$$

We say that C is the *superposition of A with B* if drawing all lengths from A and all lengths from B in the order given by A and B onto a single common line and then taking all cut positions from both lists together, yields just the cut positions of the lengths in C arranged in the order given by C.

DOUBLE DIGEST
Given three lists A, B, C of lengths, find all permutations A^*, B^*, C^* of A, B, C such that C^* is a superposition of A^* with B^*.

Consider the following special case of the DOUBLE DIGEST problem. List A contains lengths which sum up to an even number s, list B contains two identical numbers $s/2$, and list C is the same as list A. Then permutations A^*, B^*, C^* exist such that C^* is a superposition of A^* with B^* if and only if the numbers in A can be separated into two sublists with identical summed lengths $s/2$. Only under this condition it is guaranteed that the cut position $s/2$ inserted from list B does not introduce a new number into C^*. As it is well-known in complexity theory, this innocent looking problem of separating a list into two sublists with identical sums is an NP-hard problem, known as the PARTITION problem. Having thus reduced the PARTITION problem to the DOUBLE DIGEST problem shows that the latter one must be NP-hard, too. Arguments like these, i.e. reduction of a problem that is known to be NP-hard to the problem under consideration, will be studied in Chap. 5 in great detail.

2.3 Shotgun Sequencing and Shortest Common Superstrings

Shotgun sequencing leads to a set of overlapping fragments of an unknown DNA string S. Thus S contains all fragments as substrings. We say that S is a common superstring for the considered fragments. It seems to be plausible to assume that S usually is a shortest common superstring for the fragments under consideration. Of course, there are exceptions to this assumption. For example, any string S that consists of repetitions of the same character, or any string that exhibits strong periodicities leads to fragment sets having considerably shorter common superstrings than S. For realistic chromosome strings we expect the rule to hold.

Considering fragment sets, we can always delete any fragment that is a substring of another fragment. Having simplified the fragment sets this way we get so-called substring-free fragment sets. Thus we obtain the following algorithmic problem (see [28], [1] and [39]).

> **SHORTEST COMMON SUPERSTRING**
> Given a substring-free set of strings, find a shortest common superstring.

Consider strings AGGT, GGTC, GTGG. Their overlaps and overlaps lengths are shown in Fig. 2.3.

	AGGT	GGTC	GTGG
AGGT	ε	GGT	GT
GGTC	ε	ε	ε
GTGG	ε	GG	G

	AGGT	GGTC	GTGG
AGGT	0	3	2
GGTC	0	0	0
GTGG	0	2	1

Fig. 2.3. Pairwise overlaps and overlap lengths

The simplest idea for the construction of a common superstring is surely a greedy approach which always replaces two fragments $S = XY$ and $T = YZ$ having longest overlap Y with string XYZ. Let us see what happens for this example. First, AGGT and GGTC are found having longest overlap of length 3. They are replaced by AGGTC. Now AGGTC and GTGG as well as GTGG and AGGTC have only zero length overlap, thus the final common superstring is AGGTCGTGG having length 9. A "less greedy" strategy works better: first replace AGGT and GTGG having overlap of length 2 with AGGTGG, then replace AGGTGG and GGTC having overlap of length 2 with AGGTGGTC. Thus a shorter common superstring of length 8 is obtained.

Failure of a greedy approach is often an indication of the hardness of a problem. Indeed, in Chap. 5 we show that SHORTEST COMMON SUPERSTRING is an NP-hard problem, thus there is no polynomial time algorithm solving it (unless P = NP). In Chap. 6, a polynomial time algorithm is presented that returns a common superstring that is at most four times longer than a shortest common superstring. Such an algorithm is called a 4-approximation algorithm.

Finally, let us compare SHORTEST COMMON SUPERSTRING with the similar looking problem asking for a longest common substring for a given list of strings.

LONGEST COMMON SUBSTRING
Given a set of strings, find a longest common substring.

An application would be, for example, to find a longest common substring for two doctoral theses (in order to detect plagiarism). As we will see, this latter problem is efficiently solvable. It is rather obvious why both problems differ so radically from a complexity theory point of view. Whereas the former problem has to search in the unrestricted space of arbitrary superstrings, the latter problem has a search space that is considerably restricted by the given strings: only substrings are admitted. There is not a completely trivial solution for this problem. The data structure of suffix trees will help to develop a particularly efficient algorithm.

Returning to the former NP-complete problem we show how to embed it into more familiar algorithmic fields, namely Hamiltonian circuit problems, which will also lead us to the announced 4-approximation algorithm in Chap. 6. Let a substring-free list of strings $F_1, \ldots, F_m$ be given. Imagine S is a shortest common superstring for $F_1, \ldots, F_m$. We arrange strings $F_1, \ldots, F_m$ as a permutation $F_{\pi(1)}, F_{\pi(2)}, \ldots, F_{\pi(m)}$ with increasing start positions within S and get a situation that typically looks as in Fig. 2.4 (indicated for $m = 5$ strings).

Here, the following must be true:

- First string starts at the beginning of string S.
- Last strings ends at the end of string S.
- Next string starts within previous string or at the end of previous string, i.e. there are no gaps between consecutive strings.
- Next string ends strictly right of the end of previous string (set is substring-free).
- Any two consecutive strings have maximum overlap.

This is true since any deviation from these rules would allow us to construct a shorter common superstring by shifting a couple of consecutive strings to the left. Note that at the end of the permutations of strings the first used string appears again (and ends strictly right of the end of S).

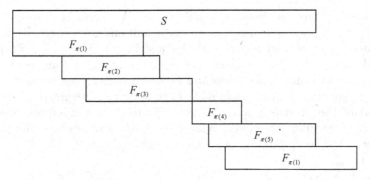

Fig. 2.4. Arrangement of fragments according to increasing start position within a common superstring S

Thus, any shortest superstring of $F_1, \ldots, F_m$ gives rise to a permutation $F_{\pi(1)}, \ldots, F_{\pi(m)}$ of strings, and conversely, any permutation $F_{\pi(1)}, \ldots, F_{\pi(m)}$ of strings gives rise to a common superstring $S(\pi)$ of $F_1, \ldots, F_m$ (not necessarily of minimal length).

$$S(\pi) = P(\pi)O(\pi) \text{ with}$$
$$P(\pi) = \text{prefix}(F_{\pi(1)}, F_{\pi(2)}) \cdots \text{prefix}(F_{\pi(m-1)}, F_{\pi(m)}) \text{prefix}(F_{\pi(m)}, F_{\pi(1)})$$
$$O(\pi) = \text{overlap}(F_{\pi(m)}, F_{\pi(1)})$$

Here we used the notions of 'prefix' and 'overlap' defined as follows. Given decompositions of strings $S = XY$ and $T = YZ$ such that Y is the longest suffix of S that is different from S and also a prefix of T, we call Y the overlap of S with T and denote it by overlap(S, T). We refer to X as prefix(S, T) and to Z as suffix(S, T) (see Fig. 2.5). Note that we require strict overlaps, i.e. an overlap is not allowed to cover the complete string. This is automatically true for different strings S and T since it was assumed that neither one is a substring of the other, but must be explicitly demanded for the overlap of a string S with itself. Thus, the overlap of string AAAA with AAAA is AAA, but not AAAA. To define the overlap of a string with itself as a strict overlap will prove to be important later.

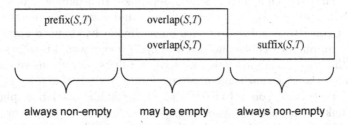

Fig. 2.5. Prefix, overlap, suffix

Besides overlaps, prefixes, and suffixes, we also consider corresponding lengths o(S, T), p(S, T), and s(S, T). These definitions give rise to two graphs, called *overlap graph* and *prefix graph*, for a string list $F_1, \ldots, F_m$. Both are directed graphs that have m nodes labelled $F_1, \ldots, F_m$ and directed edges between any two such nodes (thus also from every node back to itself). Furthermore, the edge pointing from node F_i to node F_j is weighted by number o(F_i, F_j) in the overlap graph, and p(F_i, F_j) in the prefix graph. Shortest common superstrings are closely related to cheapest cycles in prefix graph. Looking at the decomposition of string $S(\pi)$ above, we infer:

$$|S| = \sum_{i=1}^{m-1} p(F_{\pi(i)}, F_{\pi(i+1)}) + p(F_{\pi(m)}, F_{\pi(1)}) + o(F_{\pi(m)}, F_{\pi(1)})$$

$$= \text{prefixlengths}(\pi) + o(F_{\pi(m)}, F_{\pi(1)}).$$

(2.3)

Here, the term prefixlengths(π) computes the costs of the cycle through prefix graph defined by permutation π:

$$\text{prefixlengths}(\pi) = \sum_{i=1}^{m-1} p(F_{\pi(i)}, F_{\pi(i+1)}) + p(F_{\pi(m)}, F_{\pi(1)}). \qquad (2.4)$$

It is a lower bound for the length of a shortest common superstring. Note that in realistic situations with chromosome length of some 100.000 base pairs and fragments lengths of about 1000 base pairs, prefix costs of a permutation deviate only by a small fraction from the length of the superstring defined by the permutation. Thus, searching for a shortest common superstring might as well be replaced with searching for a cheapest Hamiltonian cycle (closed path visiting each node exactly once) through the prefix graph. Unfortunately, the HAMILTONIAN CYCLE problem is NP-complete, as is shown in Chap. 5. Nevertheless, it opens way for a further, computationally feasible relaxation of the concept of shortest common superstrings. As it is known from algorithm theory, computing a finite set of disjoint cycles (instead of a single cycle) having minimum summed costs and covering every node in a weighted graph is an efficiently solvable problem. Such a finite set of cycles is called a *cycle cover*. We postpone the question of how to compute a cheapest cycle cover.

At the moment let us point to a possible misunderstanding of the concept of cheapest cycle covers. Usually, weighted graphs do not have self-links from a node back to itself, equivalently stated, self-links are usually weighted with a value 0. As a consequence, in such graphs we could always consider the trivial cycle cover consisting of 1-node cycles only. Obviously, summed costs would be 0, thus minimal. Of course, such trivial cycle covers would be of no use for anything. Things change whenever we have self-links that are more costly. It could well be the case that some non-trivial path back to a node x is cheaper than the 1-node cycle from x to itself. Indeed, prefix and overlap graphs never have self-links with weight 0 (as overlaps were defined to be proper overlaps). This requirement will later prove to be essential in the construction of an approximation algorithm.

2.4 Exact Pattern Matching

2.4.1 Naive Pattern Matching

One of the most frequently executed tasks in string processing is searching for a pattern in a text. Let us start with string $T = T[1\ldots n]$ of length n, called text, and string $P = P[1\ldots m]$ of length m, called pattern. The simplest algorithm compares P with the substring of T starting at all possible positions i with $1 \leq i \leq n-m+1$. Whenever comparison was successful, start position i is returned. Obviously, this procedure requires $O(nm)$ comparisons in the worst case.

```
for i = 1 to n − m + 1 do
   a = 0
   while P(a + 1) = T(i + a) and a < m do
      a = a + 1
   end while
   if a = m then
      print i
   end if
end for
```

2.4.2 Knuth-Morris-Pratt Algorithm

There is a more clever algorithm, the so-called *Knuth-Morris-Pratt algorithm* (described in [44]), which achieves running time of $O(n + m)$ by suitably preprocessing the pattern P. Preprocessing of P works as follows. For every $a = 1, \ldots, m$ determine the length $f(a)$ of the longest proper prefix of $P[1\ldots a]$ that is also a suffix of $P[1\ldots a]$. As an example, all values $f(a)$ for pattern $P =$'ananas' are shown in Fig. 2.6.

a	1	2	3	4	5	6
$f(a)$	0	0	1	2	3	0

Fig. 2.6. Prefix-suffix match lengths computed for all prefixes of text 'ananas'

With function f available, pattern searching within text T may be accelerated as follows: start searching for P at the beginning of text T. Whenever comparison of P with T fails at some working position i of T after having successfully matched the first a characters of P with the a characters of T left of working position i, we need not fall back by a positions to start index $i - a + 1$ within T and begin comparison with P from scratch. Instead we use that $P[1\ldots f(a)]$ is the longest proper prefix of $P[1\ldots a]$ that appears left of position i within text T. Thus we may continue comparison of characters with $P(f(a)+1)$ and $T(i)$. In the pseudo-code below, variables i and a maintain the actual working position within T and the length of the actual prefix of P that was found as a substring of T exactly left of working position i. Both working position i within T as well as actual start position $i-a$ of the matching prefix of P of length a monotonically grow from 1 to at most n. Thus, we find one or all occurrences of P in T in $2n$ steps.

```
a = 0
for i = 1 to n do
    while a > 0 and P(a + 1) ≠ T(i) do
        a = f(a)
    end while
    if P(a + 1) = T(i) then
        a = a + 1
    end if
    if a = m then
        print  i − m + 1
        a = f(m)
    end if
end for
```

Figure 2.7 shows a run of the algorithm which requires two shifts to the right of pattern P to find a match with actual character x.

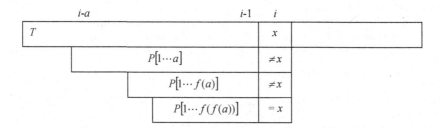

Fig. 2.7. Shifting prefixes

It remains to be clarified how expensive the computation of all values of function f is. This is done the same way by shifting prefixes of P over P, instead of shifting prefixes of P over T. The reader may develop a corresponding diagram as above.

```
f(1) = 0
a = 0
for j = 2 to m do
```

while $a > 0$ and $P(a + 1) \neq P(j)$ **do**
$\quad a = f(a)$
end while
if $P(a + 1) = P(j)$ **then**
$\quad a = a + 1$
$\quad f(j) = a$
else
$\quad f(j) = 0$
end if
end for

As with the former algorithm the whole procedure requires at most $2m$ moves of start and end index of the actual prefix of P, thus requires $O(m)$ steps. All together, $O(n + m)$ is the complexity of applying the Knuth-Morris-Pratt algorithm to a text of length n and pattern of length m.

2.4.3 Multi-Text Search

Preprocessing of search patterns is particularly valuable in case of searching for a fixed pattern of length m in several (say p) texts of length n each. Assume that $m < n$. For this case we get a running time of $O(m+pn) = O(pn)$ instead of $O(pnm)$ as for the naive algorithm. Thus we have achieved a speed-up factor of m.

2.4.4 Multi-Pattern Search

Unfortunately, in bioinformatics one usually has a fixed text of length n (say a human genome) and several (say p) patterns of length m. The Knuth-Morris-Pratt algorithm requires p times a preprocessing of a pattern of length m, followed by p times running through text T. This leads to a complexity of $O(pm + pn)$. Using suffix trees this can be done better. Construction of a suffix tree can be performed in time $O(n)$ (as shown in Chap. 4). Searching for a pattern P of length m within a suffix tree can then be done in $O(m)$ steps by simply navigating into the tree along the characters of P (for as long as possible). All together, searching for p patterns of length m each in a text of length n can be completed in time $O(n + pm)$. Applications of suffix trees are usually straight-forward, whereas efficient construction of suffix trees requires much more effort. A naive algorithm for the construction of the suffix tree for a string $T[1 \ldots n]$ of length n would first create a single link labelled with suffix $T[1 \ldots n]$, then integrate the next suffix $T[2 \ldots n]$ into the initial tree by navigating along the characters of $T[2 \ldots n]$ and eventually splitting an edge, then integrating suffix $T[3 \ldots n]$ by the same way, etc. In the worst case, $n + (n - 1) + (n - 2) + \ldots + 2 + 1$ comparisons of characters have to be done leading to a quadratic time algorithm. It requires clever ideas to save a lot of time such that a linear time algorithm results.

2.4.5 Formal Definition of Suffix Trees

The reader may use the 'ananas' example from Sect. 1.2.2 as an illustration for the formal concepts introduced here (see Fig. 1.7). Given string T of length n, a *suffix tree* for T is a rooted tree with leaves each labelled with a certain number, edges labelled with non-empty substrings of T in a certain way, additional links between inner nodes called *suffix links*, and a distinguished link to a node called *working position*, such that several properties are fulfilled. To state these properties, we introduce the notion of a *position in the tree* and the *path label* of a position. First, a position is a pointer pointing either to a node of the tree or pointing between two consecutive characters of an edge label. Given a position, its *path label* is the string obtained by concatenating all characters occurring at edges on the path from the root to the position under consideration. As a special case, the *node label* of some node is defined to be the path label of the position pointing to that node. Now, properties of a suffix tree can be defined as follows:

- Every inner node has at least two successor nodes.
- Edges leaving some node are labelled with substrings that have different first characters.
- Every leaf has as node label of some suffix $T[j \ldots n]$ of T and, in that case, is labelled with 'j'.
- Every suffix appears as a path label of a position in the tree.
- Every inner node with path label xw, for a character x and string w, has a so-called *suffix link* starting at that node and pointing to an inner node with suffix link w in case that w is non-empty, and pointing to the root otherwise.

For more extended trees, it is convenient to present them in a horizontal left-to-right manner. The suffix tree shown in Fig. 1.7 for string $T =$ 'ananas' is special in the sense that all suffixes of T appear as path labels of leaves. The following diagram shows a suffix tree[1] for the string 'mama' (Fig. 2.8).

Observe that two of its suffixes ('ma' and 'a') are represented as path labels of non-leaf positions. The reason for this is that string 'mama' contains suffixes that are at the same time prefixes of other suffixes. This is not the case for the string 'ananas'. By using an additional end marker symbol we can always enforce that suffixes are represented as path labels of leaves. Figure 2.8 shows a suffix tree for string 'mama', as well as for 'mama$'.

---

[1] In the literature, "suffix tree" for a string T is sometimes understood as suffix tree (in our sense) for string $T\$$ with additional end marker symbol, whereas a suffix tree (in our sense) is called "implicit suffix tree". Sometimes, suffix links and actual working position are not introduced as explicit components of a suffix tree. Since in the description of Ukkonen's algorithm in Chap. 4 suffix trees for prefixes of a string $T\$$ play a central role, and such prefixes do not end with end marker $\$$, we prefer to use the definition of suffix trees as given above.

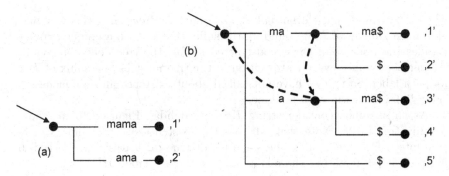

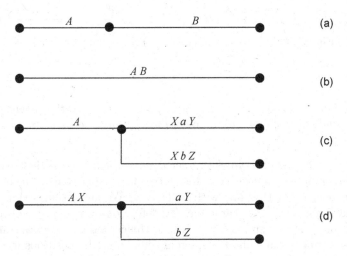

Fig. 2.8. Suffix tree: **(a)** for string 'mama'; **(b)** for string 'mama$'

Suffix trees represent all suffixes of a string T as labels of positions in a maximally compressed manner. Note that an inner node with only one successor is not allowed to occur within a suffix tree (Fig. 2.9 (a)). Instead of this, substring AB must be the label of a single edge (Fig. 2.9 (b)). Also note that a node is not allowed to have two or more edges to successors whose labels have a common non-empty maximal prefix X (Fig. 2.9 (c)). Instead of this, the situation must be represented as in Fig. 2.9 (d).

Fig. 2.9. Non-redundant labelling of edges of a suffix tree

As a consequence, a suffix tree for string $T = T[1\ldots n]$ is, up to the actual working position (that can be chosen at an arbitrary position in the tree), uniquely determined: start with a single edge with label T between a root node and a leaf marked '1'. Next, navigate along the characters of the suffix $T = T[2\ldots n]$ into the tree constructed so far, starting at the root. The restrictions on the structure of a suffix tree uniquely determine whether and where a new node must eventually be introduced. Proceeding this way with all further suffixes of T, we build up the suffix tree for T in a unique manner. Besides the nodes and edges constructed so far, the tree does not contain further nodes since every leaf position in the tree must have a suffix of T as its path label. So far, we have not talked about existence and uniqueness of suffix links. This will soon be done.

As an example, consider string $T = $ 'abcdabda'. Figure 2.10 (a) shows a suffix tree for T. Note that whereas the existence of suffix links can be guaranteed for inner nodes, the same is not true for leaves: leaf marked '6' has path label 'bda', but there is no node with path label 'da'.

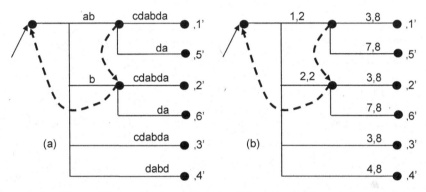

Fig. 2.10. (a) Strings as labels; (b) String limits as labels

Lemma 2.1.

For every inner node of a suffix tree with path label xw (with x being a single character and w being a string) there is a unique inner node or root with label w.

Proof. Let xw be the path label of inner node u. Each inner node has at least two successor nodes. So we may choose two leaves below node u, one with path label xwy, the other one with label xwz. We know that strings y and z must have different first characters. By definition of a suffix tree, we know that xwy and xwz are suffixes of string T. Hence, wy and wz are suffixes of T, too. By definition of suffix trees, again, there must be positions in the tree with path labels wy and wz. This means that on the path from the root down to these positions there must be a node v with path label w (at node v, the path splits into the two considered paths with labels wy and wz). Of course, node v is uniquely determined. $\qquad\square$

At first sight, it seems to be impossible to have a linear time algorithm for constructing suffix trees, since already the space requirements prevent this: for a text T of length n, a suffix tree for $T\$$ has $n+1$ leaves with concatenated labels along the paths from the root to its leaves of length $n+1, n, n-1, \ldots, 2, 1$.

So the explicit insertion of these path labels already requires $O(n^2)$ time. This problem is not too severe and can be overcome by representing an edge label $T[a \ldots b]$ by its start and end positions a and b within T, instead of explicitly writing the symbols of $T[a \ldots b]$ at the considered edge. In the example above, the suffix tree would be drawn as in Fig. 2.10 (b). Nevertheless, for better readability, in examples we prefer the string representations of edge labels. But even with such sparse representation of edge labels, $O(n^2)$ time for a naive construction of suffix trees that integrates more and more suffixes into a growing tree seems to be unavoidable since for suffixes of length $n-1, n-2, \ldots$ we must navigate in the growing tree in order to find the correct point of insertion for the actual suffix, with each navigation requiring as many steps as the number of symbols in the actual suffix.

The basic idea that will finally lead to a more efficient construction of the suffix tree of a text $T = T[1 \ldots n]$ is to successively construct suffix trees for growing strings $T = T[1 \ldots i]$. As it will be shown in Chap. 4, a suffix tree for $T = T[1 \ldots i+1]$ can be simply obtained from a suffix tree for $T = T[1 \ldots i]$ by inserting the next character $T(i+1)$ at all positions where it has to be placed. Though at first sight is seems to be a procedure that requires $O(1 + 2^2 + 3^2 + \ldots + n^2) = O(n^3)$ steps (if being implemented in a comparable naive manner as described above), we will show that most of the steps can be saved by a clever analysis of the procedure. A linear time algorithm due to Ukkonnen (see [71]) will be presented in Chap. 4. Previously published, slightly different linear time algorithms constructing suffix-trees are due to Weiner [79] and McCreight [55].

2.5 Soft Pattern Matching = Alignment

2.5.1 Alignments Restated

As introduced in Chap. 1, an alignment of string S with string T is obtained by introducing spacing symbols at certain positions of S and T such that two strings S^* and T^* of the same length result and no position in S^* and T^* is filled with two spacing symbols. As an example,

$$A\,C - G\,A - G\,T\,T\,C - A\,C\,T$$
$$- C\,T\,G\,G\,C\,T - T\,G\,G\,A - T$$

is an alignment of ACGAGTTCACT with CTGGCTTGGAT.

2.5.2 Evaluating Alignments by Scoring Functions

Quality of an alignment is measured by a numerical value called the *score* that depends on a *scoring function* σ. The scoring function expresses in terms of numerical values $\sigma(x,x)$, $\sigma(x,y)$, $\sigma(x,-)$, $\sigma(-,y)$ for different letters x and y how strongly matches x/x are rewarded and mutations x/y, deletes $x/-$ and inserts $-/y$ are punished. A typical example of a scoring function is shown in Fig. 2.11. Here, matches are rewarded with value +2, mutations are punished with value -1, inserts and deletes are even more strongly punished with value -2.

	A	C	G	T	-
A	2	-1	-1	-1	-2
C	-1	2	-1	-1	-2
G	-1	-1	2	-1	-2
T	-1	-1	-1	2	-2
-	-2	-2	-2	-2	0

Fig. 2.11. Scoring matrix

The score of an alignment S^*, T^* of string S with string T is the sum of scores of all pairs that occur at the same position in the alignment and is denoted by $\sigma^*(S^*, T^*)$. In case that S^* and T^* have common length n the definition is:

$$\sigma^*(S^*, T^*) = \sum_{i=1}^{n} \sigma(S^*(i), T^*(i)). \qquad (2.5)$$

As discussed in Chap. 1, for practically used scoring functions based on log-odds ratios of the frequencies of occurrence of character pairs and single characters, a greater score represents a more favourable event. Thus the primary goal in aligning strings is to find an alignment S^*, T^* for strings S, T having maximum score denoted as in the definition below:

$$\sigma_{\text{opt}}(S, T) = \max_{S^*, T^*} \sigma^*(S^*, T^*). \qquad (2.6)$$

Summarizing, the alignment problem is stated as follows:

ALIGNMENT
Given strings S and T, find an alignment S^*, T^* with maximum score $\sigma_{\text{opt}}(S, T)$.

Having strings S and T of length n and m, there are more than $O(2^n)$ alignments since there are already $O(2^n)$ many choices where to introduce spacing symbols into string S. The reader may think about a better lower bound for the number of alignments and show that this number is considerably greater than 2^n. The reader may also give an upper bound for the number of alignments of S and T using the fact that alignments may be at most $n + m$ pairs long.

2.6 Multiple Alignment

2.6.1 Sum-of-Pairs Maximization

Given strings $S_1, \ldots, S_k$, a multiple alignment consists of strings $T_1, \ldots, T_k$ of the same length that result from $S_1, \ldots, S_k$ by insertion of spacing symbols at certain positions, with the restriction that columns consisting of spacing symbols only are not allowed. Given a multiple alignment $T_1, \ldots, T_k$ of length m and a scoring function σ for pairs of characters $\sigma(a, b)$ and pairs of character and spacing symbol $\sigma(a, -)$ and $\sigma(-, b)$, we define its *sum-of-pairs score* as follows:

$$\sigma^*(T_1, \ldots, T_k) = \sum_{i<j} \sum_{p=1}^{m} \sigma(T_i(p), T_j(p)) = \sum_{i<j} \sigma^*(T_i, T_j). \qquad (2.7)$$

SUM-OF-PAIRS MULTIPLE ALIGNMENT
Given a scoring function σ and strings $S_1, \ldots, S_k$, compute a multiple alignment $T_1, \ldots, T_k$ with maximum sum-of-pairs score.

Computing a multiple alignment is NP-hard (shown in Chap. 5). We present a 2-approximation algorithm in Chap. 6. In the literature one finds criticism of the adequateness of scoring via sum-of-pairs. For example, from a statistical point of view, it seems quite unnatural to score a column with entries a, b, c by a sum of pairwise local log-odd values

$$\log \frac{p_{ab}}{p_a p_b} + \log \frac{p_{ac}}{p_a p_c} + \log \frac{p_{bc}}{p_b p_c} \qquad (2.8)$$

instead of taking a more integrated view expressed by common log-odds value

$$\log \frac{p_{abc}}{p_a p_b p_c}. \qquad (2.9)$$

This motivates consideration of various "consensus approaches".

2.6.2 Multiple Alignment Along a Guide Tree

The problem with sum-of-pairs alignment is that the sum of scores between any two strings of an alignment must be maximized. The problem turns into a solvable one if we consider a tree structure on the set of string indices and look for an alignment that maximizes the sum of scores between any two strings with indices connected by a link of the tree. Following the links of the tree in arbitrary order we can incrementally build up an alignment by locally maximizing scores between two linked strings. We discuss an example. Consider the tree in Fig. 2.12.

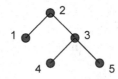

Fig. 2.12. Simple example guide tree

Assume that we have already build up an alignment T_1, T_2, T_3, T_4 with maximum sum $\sigma(T_1, T_2) + \sigma(T_2, T_3) + \sigma(T_3, T_4)$. We have to integrate a string T_5 into this alignment and take care to maximize $\sigma(T_3, T_5)$, letting the sum obtained at the stage before unchanged. Thus we should simply take a maximum score alignment T^3, T_5 of string S_3 with string S_5. The only problem is that string T^3 may have spacing symbols at different positions than the formerly computed string T_3, though both are derived from the same string S_3 by insertion of spacing symbols. A simple adaptation of T_3 and T^3 by insertion of additional spacing symbols ("padding") solves the problem. We illustrate what has to be done with a simple example:

$$
\begin{array}{ll}
T_1 & \text{C G G – T C – – G G T} \\
T_2 & \text{– G G T T – A A A G T} \\
T_4 & \text{C – – T T C – A A – G} \\
T_3 & \text{C G G T – C A A – G –}
\end{array}
$$

$$
\begin{array}{ll}
T^3 & \text{– C G G T C – A – A G} \\
S_5 & \text{A – – G T – C A T A –}
\end{array}
$$

The padded alignment adapts T_3 and T^3 without changing scores. It looks as follows.

$$
\begin{array}{ll}
R_1 & \text{– C G G – T C – – – – G G T} \\
R_2 & \text{– – G G T T – – A – A A G T} \\
R_4 & \text{– C – – T T C – – – A A – G} \\
R_3 & \text{– C G G T – C – A – A – G –}
\end{array}
$$

$$
\begin{array}{ll}
R^3 & \text{– C G G T – C – A – A – G –} \\
R_5 & \text{A – – G T – – C A T A – – –}
\end{array}
$$

We end with the following alignment:

$$
\begin{array}{ll}
R_1 & \text{– C G G – T C – – – – G G T} \\
R_2 & \text{– – G G T T – – A – A A G T} \\
R_4 & \text{– C – – T T C – – – A A – G} \\
R_3 & \text{– C G G T – C – A – A – G –} \\
R_5 & \text{A – – G T – – C A T A – – –}
\end{array}
$$

There are various heuristics that propose a more or less plausible guide tree and then align along the chosen tree. We soon discuss a few of them. Summarizing, having a tree with a distribution of strings $S_1, \ldots, S_k$ to its nodes (every node is labelled with one of the strings to be aligned) and seeking to maximize the sum of scores between strings linked in the tree makes things easy. Things change towards NP-completeness as soon as tree structures are considered that also have nodes not labelled with any one of the strings $S_1, \ldots, S_k$. This happens already for the simplest example of a tree with unlabelled root (representing a common unknown ancestor) and strings $S_1, \ldots, S_k$ attached to the leaves. Situations like these will be discussed in the next sections.

2.6.3 Consensus Line Optimization

Given a multiple alignment $T_1, \ldots, T_k$ of length m, its consensus line is the string T such that the following consensus sum has maximum value:

$$\text{consensus}(T, T_1, \ldots, T_k) = \sum_{p=1}^{m} \sum_{i=1}^{k} \sigma(T(p), T_i(p)) = \sum_{i=1}^{k} \sigma^*(T, T_i). \quad (2.10)$$

If there are several choices for a character of T, fix one in an arbitrary manner.

CONSENSUS MULTIPLE ALIGNMENT
Given a scoring function σ and strings $S_1, \ldots, S_k$, compute a multiple alignment $T_1, \ldots, T_k$ such that its consensus line T has maximum value $\text{consensus}(T, T_1, \ldots, T_k)$.

2.6.4 Steiner String

An equivalent formulation can be given exclusively in terms of strings. A *Steiner string* for a list of strings $S_1, \ldots, S_k$ is a string S that maximizes the following *Steiner sum*:

$$\text{steiner}(S, S_1, \ldots, S_k) = \sum_{i=1}^{k} \sigma_{\text{opt}}(S, S_i). \quad (2.11)$$

STEINER STRING
Given a scoring function σ and strings $S_1, \ldots, S_k$, compute a string S with maximum value $\text{steiner}(S, S_1, \ldots, S_k)$.

2.6.5 Equivalence of Consensus Lines and Steiner Strings

Theorem 2.2.
(a) From an arbitrary multiple alignment $T_1, \ldots, T_k$ of $S_1, \ldots, S_k$ with consensus line T we may construct a string S with $\text{steiner}(S, S_1, \ldots, S_k) \geq \text{consensus}(T, T_1, \ldots, T_k)$.

(b) From an arbitrary string S for $S_1, \ldots, S_k$ we may construct a multiple alignment $T_1, \ldots, T_k$ for $S_1, \ldots, S_k$ with consensus line T and
consensus$(T, T_1, \ldots, T_k) \geq$ steiner$(S, S_1, \ldots, S_k)$.
(c) If the multiple alignment in (a) has maximum consensus value of its consensus line, we obtain a Steiner string.
(d) If string S in (b) was a Steiner string, we obtain a multiple alignment with maximum consensus value of its consensus line.

Proof. (a) Consider consensus line T. Delete all occurrences of the spacing symbol from T obtaining string S. As T, T_i is an alignment (not necessarily optimal) of S with S_i we conclude that

$$\sigma_{\mathrm{opt}}(S, S_i) \geq \sigma^*(T, T_i).$$

Thus we obtain

$$\mathrm{steiner}(S, S_1, \ldots, S_k) = \sum_{i=1}^{k} \sigma_{\mathrm{opt}}(S, S_i)$$

$$\geq \sum_{i=1}^{k} \sigma^*(T, T_i)$$

$$= \mathrm{consensus}(T, T_1, \ldots, T_k).$$

(b) Consider an arbitrary string S. For each i take a maximum score alignment T^i, T_i of S with S_i. Note that though each string T^i is S with some spacing symbols introduced, spacing symbols may be placed at different positions within different strings T^i. Nevertheless, we may construct a single alignment $R_1, \ldots, R_k, R$ of $S_1, \ldots, S_k, S$ out of these local alignments by padding strings with additional spacing symbols in such a way that a common string R instead of k different strings $T^1, \ldots, T^k$ is used to align S to the other strings. Scores are not changed by such paddings. By construction, the consensus line C of alignment $R_1, \ldots, R_k$ has consensus value at least as good as R, thus

$$\mathrm{consensus}(C, R_1, \ldots, R_k) \geq \mathrm{consensus}(R, R_1, \ldots, R_k)$$

$$= \mathrm{steiner}(S, S_1, \ldots, S_k).$$

(c) Assume that we started in (a) with a multiple alignment $T_1, \ldots, T_k$ of $S_1, \ldots, S_k$ with consensus line T and maximum value consensus$(T, T_1, \ldots, T_k)$. Let S be obtained by deleting all spacing symbols from T. We know that

$$\mathrm{steiner}(S, S_1, \ldots, S_k) \geq \mathrm{consensus}(T, T_1, \ldots, T_k).$$

Now we also consider a Steiner string S_{steiner} for $S_1, \ldots, S_k$. By definition we know that

$$\mathrm{steiner}(S_{\mathrm{steiner}}, S_1, \ldots, S_k) \geq \mathrm{steiner}(S, S_1, \ldots, S_k).$$

Applying (b) to string S_{steiner} we get a further multiple alignment $R_1, \ldots, R_k$ for $S_1, \ldots, S_k$ with consensus line R and

$$\text{consensus}(R, R_1, \ldots, R_k) \geq \text{steiner}(S_{\text{steiner}}, S_1, \ldots, S_k).$$

As the initial multiple alignment was optimal we also know that

$$\text{consensus}(T, T_1, \ldots, T_k) \geq \text{consensus}(R, R_1, \ldots, R_k).$$

Combining all inequalities we obtain

$$\begin{aligned}
\text{steiner}(S, S_1, \ldots, S_k) &\geq \text{consensus}(T, T_1, \ldots, T_k) \\
&\geq \text{consensus}(R, R_1, \ldots, R_k) \\
&\geq \text{steiner}(S_{\text{steiner}}, S_1, \ldots, S_k) \\
&\geq \text{steiner}(S, S_1, \ldots, S_k).
\end{aligned}$$

Having equality in all states of the chain above, we conclude that S indeed is a Steiner string.

(d) is shown in exactly the same manner as (c).   □

2.6.6 Generalization to Steiner Trees/Phylogenetic Alignment

Given a list of strings $S_1, \ldots, S_k$, a phylogenetic tree scheme $\wp$ for $S_1, \ldots, S_k$ is a rooted tree with k leaves to which $S_1, \ldots, S_k$ are assigned in a one-to-one manner. Note that in a phylogenetic tree scheme, non-leaf nodes are not labelled, so far, with strings. Stated differently, a phylogenetic tree scheme expresses some knowledge on the evolutionary history of species $S_1, \ldots, S_k$, but without fixing how ancestors exactly looked like. Given a phylogenetic tree scheme $\wp$ for strings $S_1, \ldots, S_k$, a $\wp$-*alignment* consists of an assignment of strings to all non-leaf nodes of $\wp$. For every node u (including leaves) let $\wp(u)$ denote the string attached to u. The score of a $\wp$-*alignment* is defined as the sum of all optimal alignment scores taken over set $E(\wp)$ consisting of all pairs u, v of nodes with a link between them.

$$\text{score}(\wp) = \sum_{(u,v) \in E(\wp)} \sigma_{\text{opt}}(\wp(u), \wp(v)) \qquad (2.12)$$

> **PHYLOGENETIC TREE ALIGNMENT**
> Given a scoring function σ and a phylogenetic tree scheme $\wp$ for strings $S_1, \ldots, S_k$, compute a $\wp$-*alignment* having maximum value score($\wp$).

2.6.7 Profile Alignment

All of the problems associated with multiple alignments are NP-hard as will be shown later (Chap. 5). For each of them a 2-approximation algorithm will be presented (Chap. 6). There are other approaches to multiple alignments that admit efficient algorithms. These are based on a statistical view to multiple alignments. The simplest such approach works as follows. Given a multiple alignment $T_1, \ldots, T_k$ of length m, define its profile π as the string $\pi_1, \ldots, \pi_m$ of length m consisting of the following frequency vectors.

$$\pi_p(c) = \frac{\text{number of occurences of } c \text{ in column } p}{m} \tag{2.13}$$

Thus, "character" p of the profile of a multiple alignment is the probability distribution of characters within column p of the multiple alignment. Being a string of vector characters, the notion of an alignment of a profile with a string S is well-defined. What has to be clarified is how such an alignment is scored. This is done as follows, for all p and characters x:

$$\sigma_{\text{profile,string}}(\pi_p, x) = \sum_{\text{all characters } y} \pi_p(y)\sigma(y, x)$$

$$\sigma_{\text{profile,string}}(\pi_p, -) = \sum_{\text{all characters } y} \pi_p(y)\sigma(y, -) \tag{2.14}$$

$$\sigma_{\text{profile,string}}(-, x) = \sigma(-, x).$$

STRING TO PROFILE ALIGNMENT
Given a scoring function σ, profile π, and string S, compute a maximum score string-to-profile alignment of π with S.

We can also align a profile π to another profile ρ. Scores must obviously be defined as follows.

$$\sigma_{\text{profile,profile}}(\pi_p, \rho_q) = \sum_{\text{all characters } y,z} \pi_p(y)\rho_q(z)\sigma(y, z)$$

$$\sigma_{\text{profile,profile}}(\pi_p, -) = \sum_{\text{all characters } y} \pi_p(y)\sigma(y, -) \tag{2.15}$$

$$\sigma_{\text{profile,profile}}(-, \rho_q) = \sum_{\text{all characters } z} \rho_q(z)\sigma(-, z).$$

PROFILE TO PROFILE ALIGNMENT
Given a scoring function σ and profiles π, ρ, compute a maximum score profile-to-profile alignment of π with ρ.

2.6.8 Hidden Markov Multiple Alignment

A profile of a multiple alignment can be seen as a statistical model that describes for each position (column of the alignment) the emission probabilities for all characters (including spacing symbol). Emission probabilities at different positions are independent from each other. A major generalization is to consider statistical models with local interdependencies between adjacent positions. This gives considerably more flexibility to model evolutionary rules behind the generation of multiple alignments. Best suited are Hidden Markov models. Local interdependencies between adjacent positions are modelled by using internal states with transition probabilities between them and emission probabilities for characters associated with each internal state. Thus separating an internal behaviour of a Hidden Markov model from its observable outer behaviour is the source of increased flexibility, and also opens way to fit Hidden Markov models to data. Formal definitions of Hidden Markov models will be presented in Chaps. 3 and 7.

Figure 2.13 presents a model used for the generation of multiple alignments. Only state transitions having probability greater than zero are depicted. Example values for emission probabilities are indicated within each node of the graph. Node S indicates the start state.

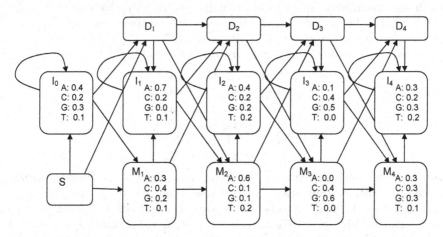

Fig. 2.13. Profile Hidden Markov model for the generation of multiple alignments

Having in mind an unknown common ancestor string of known length n ($n = 4$ in the example in Fig. 2.13) for all strings that are to be aligned (remember that an alignment partially expresses the evolutionary history of each string), the intended Hidden Markov model generates such "evolution protocols" by walking through a sequence of internal states. Available states are "match/mutate" states $M_1, \ldots, M_n$, "insert" states $I_0, I_1, \ldots, I_n$, and "delete" states $D_1, \ldots, D_n$. Walking through state M_k means to match or mutate (via character emission) the kth ancestor character, walking through state I_k means to insert fresh characters (via character emission) right of the position of the kth ancestor symbol, walking through state D_k means to delete the kth ancestor character (here no character is emitted, i.e. delete states are "mute").

We illustrate how the model works with an example. Assume there are a couple of strings to be aligned. The number of characters in the unknown common ancestor string is usually estimated to be the arithmetical mean of the lengths of the considered strings. In our example, let this be 4. Further assume that for some of the strings there already exists a plausible multiple alignment (obtained by one of the methods described before or proposed by biologists on basis of plausibility). As an example, consider the following multiple alignment.

$$
\begin{array}{ccccccccc}
1 & & & 2 & 3 & & 4 & \\
- & C & - & - & G & - & - & A & - \\
- & - & - & - & G & C & A & A & A \\
- & C & - & C & - & A & - & G & - \\
- & T & - & - & T & C & C & T & - \\
A & C & G & G & G & C & - & C & - \\
\end{array}
$$

Here, we have decided to interpret columns with a majority of characters over spacing symbols to be derived from one of the conjectured four characters of the unknown ancestor by either match, mutate, or delete events. On the other hand, columns with a majority of spacing symbols are assumed to be the result of insert events. On basis of these decisions we next introduce at all positions of the multiple alignment the correct state that led to the entry.

$$
\begin{array}{ccccccccc}
1 & & & 2 & 3 & & 4 & \\
- & M_1 & - & - & M_2 & D_3 & - & M_4 & - \\
- & D_1 & - & - & M_2 & M_3 & I_3 & M_4 & I_4 \\
- & M_1 & - & I_1 & D_2 & M_3 & - & M_4 & - \\
- & M_1 & - & - & M_2 & M_3 & I_3 & M_4 & - \\
I_0 & M_1 & I_1 & I_1 & M_2 & M_3 & - & M_4 & - \\
\end{array}
$$

Now we estimate transition and emission probabilities of the Hidden Markov model as relative frequencies. For example, there are two transitions from state M_1 to state M_2, one transition to state D_2, and one transition to state I_1. Thus, transition probabilities from state M_1 to other states are estimated as follows: 50% to M_2, 25% to D_2, 25% to I_1. Emission probabilities are estimated similarly, for example, for state M_4 emission probability is estimated for character A as 40%, for C as 20%, for G as 20%, and for T as 20%.

So far, parameters of the model have been fixed on basis of available aligned strings. The general algorithmic problem behind parameter choice is training a model or fitting a model to available data. Parameters are chosen in such a way that the resulting model M has maximum probability of generating the training data. This is called maximization of model likelihood. The required formal concepts to make this precise are the probability that model M generates data D, briefly denoted $P_M(D)$, as well as the likelihood of model M given data $D_1, \ldots, D_k$ defined by

$$
L(M|D_1, \ldots, D_k) = \prod_{i=1}^{k} P_M(D_k). \tag{2.16}
$$

Thus the following general problem is associated with every sort of adaptive system with parameters that may be optimally fitted to training data.

HIDDEN MARKOV MODEL TRAINING
Given the structure of a Hidden Markov model (i.e. number of states and connection structure, but without having fixed parameters) and observed data $D_1, \ldots, D_k$, compute model parameters such that the resulting model has maximum likelihood of generating the data.

In Chap. 3 we will discuss the *Baum-Welch algorithm* which achieves a local maximization of model likelihood. Global maximization of model likelihood is a hard to solve problem.

Proceeding with our example, assume now that a further string CACGCTC has to be integrated into the alignment above. Referring to the chosen model, a most probable state sequence can be computed by the so-called *Viterbi algorithm*.

HIDDEN MARKOV MODEL DECODING
Given a Hidden Markov model and observed data D, compute a most probable state sequence that emits D.

Details of this algorithm are postponed to Chap. 3 on dynamic programming.

Assume here that as the most probable state sequence the algorithm proposes $M_1 I_1 I_1 M_2 M_3 I_3 D_4 I_4$. Note that there are seven match/insert states corresponding to the seven characters of the string, and four match/delete states corresponding to the four characters of the unknown ancestor. The correct extension of the alignment thus looks as follows.

```
        1       2 3   4
      - C - - G - - A -
      - - - - G C A A A
      - C - C - A - G -
      - T - - T C C T -
      A C G G G C - C -
      - C A C G C T - C
```

One says that a further string "has been aligned to the Hidden Markov model".

2.6.9 Example Heuristic for Obtaining Multiple Alignments

The following incremental construction of multiple alignments is plausible and widely used. Given strings $S_1, \ldots, S_k$ we first compute for every pair of indices $i < j$ value $\sigma_{\text{opt}}(S_i, S_j)$. Then we use these values to apply a standard clustering procedure, as for example CLUSTALW, to group the most similar strings S_a and S_b into a group S_{ab}. We treat this new group S_{ab} as a single new object, define scores $\sigma_{\mathrm{opt}}(S_i, S_{ab})$ as arithmetical mean of $\sigma_{\mathrm{opt}}(S_i, S_a)$ and $\sigma_{\mathrm{opt}}(S_i, S_b)$, thus obtaining a tree with strings $S_1, \ldots, S_k$ attached to its leaves. This is a rather plausible guide tree which can be used to efficiently construct a multiple alignment.

2.7 Gene Detection

Having sequenced DNA strings the next important task is to identify functional units, e.g. genes or regulatory regions. We concentrate on gene detection. Biology contributes several indications for a substring to be part of an exon that belongs to a gene, e.g. absence of a stop codon, certain short strings that are characteristic for beginning or ending of an exon, or a higher frequency of CpG pairs in exon regions when compared to non-exon regions. Known frequencies of CpG pairs in exon versus non-exon regions, for example, can be used to probabilistically classify on a likelihood basis strings into exons versus non-exons. Such approaches, as also the more sophisticated Hidden Markov model approaches, for predicting CpG-islands are based on standard methods of probability theory. However, they do not require sophisticated algorithmic know-how and we do not further discuss them in this book. Instead we concentrate on a different approach of *gene assembly* that will later be solved by dynamic programming. In gene assembly one assumes that a new strand G of DNA has been sequenced that is suspected to contain the exons of a gene producing a certain protein T (for simplicity assuming that it is given by its DNA code). We assume that well-known heuristics for predicting exons give us an ensemble of substrings $E_1, \ldots, E_b$ of G that are suspected to be valid exons. We assume that indeed all true exons are covered by the list of candidate exons $E_1, \ldots, E_b$, whereas also non-exons may occur among this list. The task arises to select an ascending chain Γ of non-overlapping candidate exons (black shaded areas) whose concatenation Γ^* gives greater or equal optimal alignment score with T than all other selections of ascending chains (Fig. 2.14).

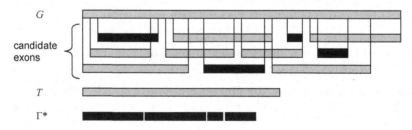

Fig. 2.14. Optimal alignment of Γ^* with T gives best value among all possible chains.

Letting T range over a database of known DNA codes of proteins, we may find one string T which can best be aligned with a concatenation of an ascending chain of candidate exons. Thus, with some confidence, we find out which protein is probably synthesized by the expected gene within string G and moreover, which substrings of G are the putative exons within that gene. Of course, simply trying out all ascending chains one by one is no good idea for obtaining an efficient algorithm. We will later see that clever usage of dynamic programming leads to an efficient solution (Chap. 3).

2.8 Genome Rearrangement

In Chap. 1 we discussed as a particular case of genome reorganization the operation of directed reversal. Encoding a sequence of n genes distributed on both strands of a chromosome as a permutation of signed numbers, i.e. numbers equipped with a sign ($+$ or $-$), a directed reversal takes a segment of the sequence, inverts ordering of numbers, as well as inverts sign for each number. As an example, signed permutation -4 +2 +6 -1 -3 +5 might be transformed by a single signed reversal into -4 +3 +1 -6 -2 +5, but also into +4 +2 +6 -1 -3 +5. Given two signed permutations of n numbers, one is interested in computing the least number of directed reversals required to transform one signed permutation into the other. This minimum number is called the (directed) *reversal distance* between two signed permutations. As can be easily verified, it defines a metric on signed permutations of n numbers. By renumbering genes and exchanging signs one can always assume that the target signed permutation is the sorted one +1 +2 ... +n. Thus the problem simplifies to the task to compute the least number of directed reversals required to transform a given signed permutation into the sorted one. The problem is therefore also called the problem of sorting a signed permutation.

SIGNED PERMUTATION SORTING
Given a signed permutation π of n numbers, compute the least number $d(\pi)$ of directed reversals required to sort permutation π.

A similar problem arises for unsigned permutations, i.e. permutations of n numbers without $+$ or $-$ sign attached to them.

UNSIGNED PERMUTATION SORTING
Given an unsigned permutation π of n numbers, compute the least number $d(\pi)$ of undirected reversals required to sort permutation π.

It will turn out that these problems, though looking quite similar, are radically different from a complexity point of view. One of them admits an efficient algorithm, the other one is NP-hard (see the series of papers [41, 42, 43] and also [15]). To give the reader a first impression of how complex the problem of sorting permutations is, sorting of a signed permutation of 36 signed numbers as shown in Fig. 2.15 requires 26 directed reversals. In Chap. 4 we will see that 26 is indeed the least number of directed reversals required for sorting the presented signed permutation.

```
-12 +31 +34 -28 -26 +17 +29 +04 +09 -36 -18 +35 +19 +01 -16 +14 +32 +33 +22 +15
-11 -27 -05 -20 +13 -30 -23 +10 +06 +03 +24 +21 +08 +25 +02 +07
+20 +05 +27 +11 -15 -22 -33 -32 -14 +16 -01 -19 -35 +18 +36 -09 -04 -29 -17 +26
+28 -34 -31 +12 +13 -30 -23 +10 +06 +03 +24 +21 +08 +25 +02 +07
+01 -16 +14 +32 +33 +22 +15 -11 -27 -05 -20 -19 -35 +18 +36 -09 -04 -29 -17 +26
+28 -34 -31 +12 +13 -30 -23 +10 +06 +03 +24 +21 +08 +25 +02 +07
+01 -16 -15 -22 -33 -32 -14 -11 -27 -05 -20 -19 -35 +18 +36 -09 -04 -29 -17 +26
+28 -34 -31 +12 +13 -30 -23 +10 +06 +03 +24 +21 +08 +25 +02 +07
+01 -16 -15 -36 -18 +35 +19 +20 +05 +27 +11 +14 +32 +33 +22 -09 -04 -29 -17 +26
+28 -34 -31 +12 +13 -30 -23 +10 +06 +03 +24 +21 +08 +25 +02 +07
+01 -16 -15 -14 -11 -27 -05 -20 -19 -35 +18 +36 +32 +33 +22 -09 -04 -29 -17 +26
+28 -34 -31 +12 +13 -30 -23 +10 +06 +03 +24 +21 +08 +25 +02 +07
+01 -16 -15 -14 +31 +34 -28 -26 +17 +29 +04 +09 -22 -33 -32 -36 -18 +35 +19 +20
+05 +27 +11 +12 +13 -30 -23 +10 +06 +03 +24 +21 +08 +25 +02 +07
+01 +26 +28 -34 -31 +14 +15 +16 +17 +29 +04 +09 -22 -33 -32 -36 -18 +35 +19 +20
+05 +27 +11 +12 +13 -30 -23 +10 +06 +03 +24 +21 +08 +25 +02 +07
+01 +26 +28 +18 +36 +32 +33 +22 -09 -04 -29 -17 -16 -15 -14 +31 +34 +35 +19 +20
+05 +27 +11 +12 +13 -30 -23 +10 +06 +03 +24 +21 +08 +25 +02 +07
+01 +26 +28 +29 +04 +09 -22 -33 -32 -36 -18 -17 -16 -15 -14 +31 +34 +35 +19 +20
+05 +27 +11 +12 +13 -30 -23 +10 +06 +03 +24 +21 +08 +25 +02 +07
+01 +26 +28 +29 +30 -13 -12 -11 -27 -05 -20 -19 -35 -34 -31 +14 +15 +16 +17 +18
+36 +32 +33 +22 -09 -04 -23 +10 +06 +03 +24 +21 +08 +25 +02 +07
+01 +26 +11 +12 +13 -30 -29 -28 -27 -05 -20 -19 -35 -34 -31 +14 +15 +16 +17 +18
+36 +32 +33 +22 -09 -04 -23 +10 +06 +03 +24 +21 +08 +25 +02 +07
+01 +26 +27 +28 +29 +30 -13 -12 -11 -05 -20 -19 -35 -34 -31 +14 +15 +16 +17
+18 +36 +32 +33 +22 -09 -04 -23 +10 +06 +03 +24 +21 +08 +25 +02 +07
+01 +26 +27 +28 +29 +30 +31 +34 +35 +19 +20 +05 +11 +12 +13 +14 +15 +16 +17 +18
+36 +32 +33 +22 -09 -04 -23 +10 +06 +03 +24 +21 +08 +25 +02 +07
+01 +26 +27 +28 +29 +30 +31 +34 +35 +19 +20 +09 -22 -33 -32 -36 -18 -17 -16 -15
-14 -13 -12 -11 -05 -04 -23 +10 +06 +03 +24 +21 +08 +25 +02 +07
+01 +26 +27 +28 +29 +30 +31 +22 -09 -20 -19 -35 -34 -33 -32 -36 -18 -17 -16 -15
-14 -13 -12 -11 -05 -04 -23 +10 +06 +03 +24 +21 +08 +25 +02 +07
+01 +26 +27 +28 +29 +30 +31 +32 +33 +34 +35 +19 +20 +09 -22 -36 -18 -17 -16 -15
-14 -13 -12 -11 -05 -04 -23 +10 +06 +03 +24 +21 +08 +25 +02 +07
+01 +26 +27 +28 +29 +30 +31 +32 +33 +34 +35 +36 +22 -09 -20 -19 -18 -17 -16 -15
-14 -13 -12 -11 -05 -04 -23 +10 +06 +03 +24 +21 +08 +25 +02 +07
+01 +26 +27 +28 +29 +30 +31 +32 +33 +34 +35 +36 +22 -09 -24 -03 -06 -10 +23 +04
+05 +11 +12 +13 +14 +15 +16 +17 +18 +19 +20 +21 +08 +25 +02 +07
+01 +26 +27 +28 +29 +30 +31 +32 +33 +34 +35 +36 +22 -09 -08 -21 -20 -19 -18 -17
-16 -15 -14 -13 -12 -11 -05 -04 -23 +10 +06 +03 +24 +25 +02 +07
+01 +26 +27 +28 +29 +30 +31 +32 +33 +34 +35 +36 +08 +09 -22 -21 -20 -19 -18 -17
-16 -15 -14 -13 -12 -11 -05 -04 -23 +10 +06 +03 +24 +25 +02 +07
+01 +26 +27 +28 +29 +30 +31 +32 +33 +34 +35 +36 +08 +09 -22 -21 -20 -19 -18 -17
-16 -15 -14 -13 -12 -11 -05 -04 -03 -06 -10 +23 +24 +25 +02 +07
+01 +26 +27 +28 +29 +30 +31 +32 +33 +34 +35 +36 +08 +09 -22 -21 -20 -19 -18 -17
-16 -15 -14 -13 -12 -11 -05 -04 -03 -02 -25 -24 -23 +10 +06 +07
+01 +02 +03 +04 +05 +11 +12 +13 +14 +15 +16 +17 +18 +19 +20 +21 +22 -09 -08 -36
-35 -34 -33 -32 -31 -30 -29 -28 -27 -26 -25 -24 -23 +10 +06 +07
+01 +02 +03 +04 +05 +11 +12 +13 +14 +15 +16 +17 +18 +19 +20 +21 +22 -09 -08 -07
-06 -10 +23 +24 +25 +26 +27 +28 +29 +30 +31 +32 +33 +34 +35 +36
+01 +02 +03 +04 +05 +06 +07 +08 +09 -22 -21 -20 -19 -18 -17 -16 -15 -14 -13 -12
-11 -10 +23 +24 +25 +26 +27 +28 +29 +30 +31 +32 +33 +34 +35 +36
+01 +02 +03 +04 +05 +06 +07 +08 +09 +10 +11 +12 +13 +14 +15 +16 +17 +18 +19 +20
+21 +22 +23 +24 +25 +26 +27 +28 +29 +30 +31 +32 +33 +34 +35 +36
```

Fig. 2.15. Sorting an extended signed permutation

2.9 RNA Structure Prediction

RNA macromolecules are versatile and have the ability to fold into diverse structures. The common scaffold comprises base pairing between complementary bases. Formally speaking, a *structure* R over an RNA sequence $S = S[1 \ldots n]$ is defined as a set of base pairs:

$$R = \{(i,j) \mid 1 \leq i < j \leq n \wedge i \text{ and } j \text{ form an admissible base pair}\}$$
$$\text{such that } \forall (i,j), (i',j') \in R : i = i' \Leftrightarrow j = j'.$$

This definition demands that in an RNA structure R, a base can form a pair with at most one other base. Additionally, one commonly asks for every base pair (i,j) in R to be separated by at least one base, i.e. $j - i \geq 2$. In more advanced prediction methods, an even longer distance between paired bases i and j is demanded to serve for naturally occuring RNA structures. Structure R is called *free of pseudoknots* if there are no two pairs (i,j) and (k,l) in P such that $i < k < j < l$. Allowing overlapping base pairs results in a pseudoknot structure (Fig. 2.16).

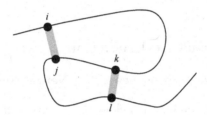

Fig. 2.16. Pseudoknot structure with overlapping base pairs

A simple approach for RNA structure prediction from sequence is based on maximizing the number of base pairs.

BASE PAIR MAXIMIZATION
Given an RNA sequence S, compute a structure R with maximum number of base pairs.

More sophisticated optimization methods attempt to minimize the so-called free energy of a fold. Out of the exponential number of possible structures, an RNA molecule will fold into the one with minimum free energy. The state of *minimum free energy* is always determined by both enthalpic and entropic forces. It is clear that by considering only stable stems, RNA folding is very much oversimplified. In RNA, enthalpic terms arise from base pairing (stabilizing forces) and entropic terms from unstructured regions (destabilizing forces). RNA comprises various *structure elements*, e.g. stems, hairpin loops, bulges, internal loops, and multiloops, which we introduced in Sect. 1.6 (a formal definition is postponed to Sect. 3.6). All these motifs contribute to the overall free energy and much experimental work has been done to determine their free (positive or negative) energy parameters. Stems act as stabilizing elements in folds, thus add a negative value to the free energy. Loop regions act as destabilizing elements in folds, expressed by a positive contribution to free energy. The result is a refined variant of the RNA structure prediction problem with the goal to minimize the free energy.

FREE ENERGY MINIMIZATION
Given an RNA sequence S, compute a structure R with minimum overall free energy.

Note that in a pseudoknot-free fold secondary structure elements occur either nested or side by side, but not interleaving. If we ignore pseudoknots, an RNA structure with maximum number of base pairs or minimum free energy can be computed in polynomial time using dynamic programming. We will introduce these successful and elegant approaches in Sect. 3.6. However, including more complicated motifs such as pseudoknots dampens the optimism of solving the RNA structure prediction problem. The general pseudoknot structure prediction problem is NP-complete and the proof for this based on [52] will be delivered in Sect. 5.5.6.

2.10 Protein Structure Prediction

2.10.1 Comparative Approaches Using Neural Networks

Predicting whether an amino acid belongs to an alpha-helix, beta-sheet or coil (loop) structure is a classical task for *feedforward neural networks* (also called *multi-layer perceptrons*). These are supplied with a certain number of amino acids surrounding the amino acid whose secondary structure class is to be predicted. Having enough such windows with known class of its centre amino acid, there is a good chance that a properly designed feedforward network may be trained to these data - and also generalise well on so far unseen data. Moving from secondary structures towards tertiary structure of proteins, an intermediate problem is contact map prediction, i.e. to predict for every pair of amino acids whether they are nearby in the natural fold of the protein. Whenever there is a contact predicted between amino acids A_i and A_j, it is reasonable to assume that both these amino acids belong to a core structure which is usually highly conserved throughout all members of a family $\wp$ of homologous proteins. This should be observable in a high positive covariance with respect to mutations at sites i and j in protein family $\wp$. Moreover, characteristic covariance patterns should also be observable between all pairs of sites in the surroundings of sites i and j. This motivates an approach that presents to a neural network, among other parameters as the above discussed secondary structure class predictions, lots of standard covariance coefficients, taken from a multiple alignment of a protein family $\wp$. Having now considerably more parameters as an input for a neural network than in the example before, taking care of the notorious problem of overfitting in neural learning becomes more urgent. Support vector machines are particularly well suited to solve this problem. Indeed, there are successful approaches to contact map prediction using support vector machines.

As the emphasis of this book lies more on traditional (exact) combinatorial algorithms, we do not further develop neural network techniques in the main chapters of this book, but postpone this theme to Chap. 7 on metaheuristics, and now switch back to a more classical combinatorial problem.

2.10.2 Protein Threading

Let S be a string of n amino acids describing the primary structure of a novel protein. Let a reference protein T of k amino acids be given with known tertiary structure consisting of a sequence of core segments (alpha or beta) $C_1, \ldots, C_m$ of fixed lengths $c_1, \ldots, c_m$ that are separated by loop segments $L_0, L_1, \ldots, L_{m-1}, L_m$ of lengths $l_0, l_1, \ldots, l_{m-1}, l_m$. Thus, reference protein T is segmented into a concatenation of substrings

$$T = L_0 C_1 L_1 C_2 \ldots C_m L_m \qquad (2.17)$$

with loop and core segments of known lengths occurring in alternation as shown in Fig. 2.17.

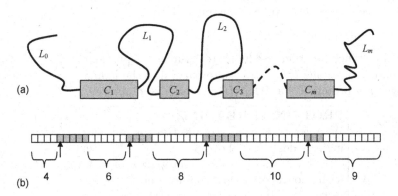

Fig. 2.17. (a) Core structure of reference protein; (b) Threading into reference protein

Assume that length c_i of core segment C_i is strongly conserved in evolution, whereas length l_i of loop segment L_i varies within an interval $[\lambda_i \ldots \Lambda_i]$. The task is to identify corresponding core segments $C^1, \ldots, C^m$ of lengths $c_1, \ldots, c_m$ within the novel protein S such that a certain distance measure between $C_1, \ldots, C_m$ and $C^1, \ldots, C^m$ is minimized and, of course, constraints on the lengths of loop segments are respected. Let core segments $C^1, \ldots, C^m$ of fixed lengths $c_1, \ldots, c_m$ within protein S be defined by their start positions $t_1, \ldots, t_m$ within reference protein T. Such a list of start positions is called a *threading* of S into T. As an example, let core lengths $5, 4, 6, 3$ and loop length intervals $[3 \ldots 7], [3 \ldots 6], [7 \ldots 9], [6 \ldots 12], [7 \ldots 11]$ be given (Fig. 2.17). Then, start positions $5, 16, 28, 44$ of the shaded core segments define an admissible threading of protein S into T. Admissibility of a threading may be expressed by the following inequalities ($i = 2, \ldots, m - 1$):

$$1 + \lambda_0 \leq t_1 \leq 1 + \Lambda_0$$
$$t_i + c_i + \lambda_i \leq t_{i+1} \leq t_i + c_i + \Lambda_i \qquad (2.18)$$
$$t_m + c_m + \lambda_m \leq n + 1 \leq t_m + c_m + \Lambda_m.$$

Using this notation, the quality of a threading is measured by a function $f(t_1, \ldots, t_m)$.

$$f(t_1, \ldots, t_m) = \sum_{i=1}^{m} g(t_i) + \sum_{i=1}^{m-1} \sum_{j=i+1}^{m} h(t_i, t_j) \qquad (2.19)$$

Here, term $g(t_i)$ is an abbreviation for some function depending on core segment C_i of T and substring $C^i(t_i)$ of S with length c_i that starts at position t_i. Furthermore, $h(t_i, t_j)$ is an abbreviation for some function depending on corresponding substrings C_i, $C^i(t_i)$, C_j, and $C^j(t_j)$. *Local term* $g(t_i)$ may be used as a measure of how good core segment C^i of S resembles core segment C_i of T, whereas *coupling term* $h(t_i, t_j)$ may be used to express interactions between core segments. An example might be

$$g(t_i) = d_{\mathrm{opt}}(C_i, C^i(t_i)) \text{ with } C^i(t_i) = S[t_i \ldots t_i + c_i - 1]$$

with a distance function d. The protein threading problem now attempts to compute a threading $t_1, \ldots, t_m$ of protein S into the loop and core structure of reference protein T such that $f(t_1, \ldots, t_m)$ is minimized.

> **PROTEIN THREADING**
> We refer to a fixed scoring $f(t)$ for any thread-ing t as described above. Given a novel protein S of length n and a reference protein T of length k with known core-loop structure, and also given in-tervals for admissible core lengths, determine an admissible threading t of S into the structure of T with minimum value $f(t)$.

In Chap. 3 we will see that this problem is efficiently solvable by a dynamic programming approach provided the evaluation function $f(t)$ does not contain any coupling terms, whereas in Chap. 5 we show that the presence of simplest coupling terms in $f(t)$ immediately leads to NP-hardness of the problem (see [50]).

2.11 Bi-Clustering

We start with a matrix of binary entries with n rows corresponding, for example, to n different microarray measurements, each involving the expression of m genes (columns of the matrix). We assume that measurements are taken in the context of a disease exhibiting a few different variants, each of them characterized by a different gene expression background. What we would like to have is a prediction algorithm telling us to which variant each measurement (row) belongs and which gene expression pattern characterizes measurements of a certain variant. Thus, this is on the one hand a problem of clustering measurements and on the other hand, of finding gene expression rules that characterize clusters. One can imagine various sorts of rules for the characterization of a cluster of measurements:

- Expression rule: characterize a cluster by a fixed subset of genes that must be expressed in measurements of the cluster.
- Expression/suppression rule: characterize a cluster by a fixed subset of genes that must be expressed and a fixed subset of genes that may not be expressed in measurements of the cluster.
- Arbitrary Boolean rule: characterize a cluster by validity of a Boolean formula on the expression/suppression states of the genes in a fixed set of genes in all measurements of the cluster.

It is expected that number of clusters as well as size of each cluster have to be restricted to make the result of clustering and characterization of clusters meaningful, for example by expecting k clusters with at least K elements within each cluster for suitably chosen numbers k, K. Also, admitted rules should be somehow restricted. Admitting, for example, every Boolean formula is surely no good idea, since it allows characterization of each cluster in an arbitrary clustering of measurements by simply using as a rule characterizing a considered cluster the Boolean formula in disjunctive normal form that simply enumerates all measurements in the cluster. We thus fall back, as an example, to the simplest sort of cluster rules by gene expression. Note that we look at the same time for clusters and gene expression rules for each cluster. This is different from the task to find, for example, for a partitioning into two classes a decision tree that separates classes. For such a situation with known clusters one could apply standard techniques of machine learning (e.g. version space procedure or ID3) for the inference of decision trees. Here, the different problem of simultaneously finding clusters and gene expression characterizations for each cluster has to be solved. As a first problem variant (the simplest one among lots of further ones) we discuss the following.

MAXIMUM SIZE ONLY-ONES SUBMATRIX
Given binary matrix M and lower bound b, is there submatrix of size at least b (size of a matrix = number of entries) that contains only entries '1'.

As will be shown in Chap. 5, already this simply looking problem is NP-complete (shown in [63]).

2.12 Bibliographic Remarks

The most comprehensive treatment of algorithmic problems resulting from bioinformatics problems is surely Gusfield [31]. Various more special, and in parts rather intricate problems are treated in Pevzner [64].

3

Dynamic Programming

3.1 General Principle of Dynamic Programming

As numerous examples from algorithm theory show, a *recursive solution* of a problem following the general principle of *divide-and-conquer* usually has the advantage of being clear and understandable, and often leads to an efficient algorithmic solution of a problem. In a recursive approach, any instance x of a problem is reduced to smaller instances $x_1, \ldots, x_k$ that are solved separately. Then, from solutions $y_1, \ldots, y_k$ of $x_1, \ldots, x_k$ a solution y of the original instance x is assembled, usually in a rather simple and cheap manner. To particularly solve an optimization problem in a recursive manner depends on the validity of the so-called *Bellman principle* (see [8], reprint [9]): assembly of an optimal solution y of x requires optimal solutions $y_1, \ldots, y_k$ of instances $x_1, \ldots, x_k$.

Unfortunately, implementing a recursive solution strategy directly by using recursive procedures often leads to exponential running time (measured in terms of input length), with the exact running time estimation depending on the number and size of the generated sub-instances and the costs for assembling sub-solutions into a solution of the original call. For example, replacement of every instance x of length n by two instances x_1 and x_2, each of length $n - 1$, obviously leads to an exponential number of recursive subcalls, thus to execution time at least 2^n, whereas replacement of every instance x of length n by two instances x_1 and x_2, each of length $n/2$ as done in good sorting algorithms, leads to the well-known execution time of $O(n \log n)$. Even in case that a direct recursive implementation leads to an exponential number of recursive subcalls, the game is not necessarily lost. It might well be the case that there are only polynomially many *different* recursive subcalls, but with multiple calls of each of them. This offers the opportunity to compute all these subcalls in a bottom-up tabular manner, starting with the smallest instances, storing computed solutions for all of them, and using these to compute solutions for longer instances, too. Stated differently, top-down recursive computation is replaced by a bottom-up iterative computation of solutions of all occurring instances, accompanied by storage of already computed solutions for usage in the computation of further solutions.

The most prominent and simplest application of dynamic programming is the computation of Fibonacci numbers. The ith Fibonacci number $f(i)$ is recursively defined by the following equations: $f(0) = 1, f(1) = 1, f(i + 2) = f(i) + f(i + 1)$. A top-down recursive execution of the recursion when calling $f(n)$ leads to at least $2^{n/2}$ many subcalls of the Fibonacci function though there are only $n - 1$ different subcalls to be executed. If these are computed iteratively in a bottom-up manner, only $n - 1$ calls of the Fibonacci function must be evaluated. Contrasting this example, we next describe a situation where dynamic programming is not applicable. Consider testing satisfiability of Boolean formulas $\varphi(x_1, \ldots, x_n)$ having n Boolean variables. Such a formula is satisfiable if and only if at least one of the formulas $\varphi(\text{'true'}, \ldots, x_n)$, $\varphi(\text{'false'}, \ldots, x_n)$ is satisfiable. This is a recursive formulation of satisfiability. Here, we end up with 2^n different subcalls. Indeed, satisfiability is an NP-hard problem so that a polynomial time solution is not possible, unless P = NP.

The main steps of any dynamic programming approach can be summarized as follows:

- *Parameterization*: think about (usually natural and canonical) ways to reduce any problem instance x to a couple of smaller sub-instances.
- *Recursive solution*: show how solutions of sub-instances may be assembled into an overall solution.
- *Bellman principle*: show that an optimal solution of any instance x must necessarily be assembled from optimal solutions of the sub-instances.
- *Counting and overall complexity*: count the number of different instances of size at most the size of x. Show that this number is bounded by a polynomial in the size of x. Estimate the costs of computing the solution of an instance x from stored solutions of its sub-instances. Together with the estimated number of sub-instances give an overall estimation of the complexity of the procedure.
- *Tabular (bottom-up) computation*: organize the computation of solutions of all sub-instances of x in a tabular manner, proceeding from smaller instances to greater instances.

There will be a couple of further tricks that are often applied in dynamic programming to improve complexity; these are illustrated in the following examples. To complete the listing above we now reveal them in advance. Sometimes, recursion requires some sort of additional constraints to be maintained. In case of a fixed (static) number of different constraints these may be incorporated by treating several optimization functions in parallel instead of a single one. In more complex cases of constraints dynamically depending on input, it may be necessary to incorporate constraints into the definition of the optimization function via additional parameters controlling the required constraints. So far, for a recursively defined function f, the bottom-up evaluation of all subcalls of f reduced running time by avoiding multiple calls of identical terms. Further reductions of running time may be achieved by treating functional terms that frequently occur in these subcalls of function f in a more efficient manner. Let us call terms without any occurrence of function f within them *basic terms*, and terms with an occurrence of f within them *recursive terms*. Concerning basic terms, computing and storing them before the main bottom-up computation of subcalls of f starts, always saves running time. Concerning recursive terms, it may be advantageous to compute them within a separate recursive procedure parallel to the main recursion, instead of nested within the main recursion. Thus, the list above may be completed as follows:

- *Conditioning*: integrate constraints that are required for a successful formulation of recursion into the definition of the optimization function by either statically using several functions instead of a single one, or by expressing dynamically varying constraints by a further function parameter.
- *Preprocessing of basic sub-terms*: look for basic terms (without recursive calls occurring within them) which may be computed in advance before starting the recursion.
- *Parallel processing of recursive sub-terms*: look for recursive terms which may be computed in parallel to the main recursion.

We illustrate these approaches by a couple of bioinformatics problems. The simplest example is *alignment* of strings. Here, dynamic programming in its basic and pure form can be studied. Integration of static constraints is required to solve *affine gap alignment*, which modifies alignment in the sense that spacing symbols (gaps) are scored in a block-wise manner instead of scoring them additively in a pointwise fashion. Integration of dynamically varying constraints is required for *gene assembly*, the computation of optimal *RNA folds*, Viterbi algorithm for *Hidden Markov models*, optimal *lifted phylogenetic alignment*, *protein threading*, and *profile alignment*.

3.2 Alignment

3.2.1 Problem Restated

Given scoring function σ and strings S, T of length n, m, we want to compute an alignment S^*, T^* of S with T having maximum score. We concentrate here on the computation of the maximum score $\sigma_{\text{opt}}(S, T) = \max_{S^*, T^*} \sigma^*(S^*, T^*)$. From this, an optimal alignment can be easily extracted by tracing back the computation leading to this maximal score. Note that simply enumerating all possible alignments results in exponential running time: although the lengths of strings S^* and T^* in an alignment of S with T can be bounded by $n + m$, there are nevertheless $O(2^{n+m})$ ways to introduce spacing symbols into strings S and T.

3.2.2 Parameterization

A natural way to define sub-instances for the optimal alignment of strings S and T is to optimally align certain substrings of S and T. It turns out that optimally aligning prefixes of S with prefixes of T is sufficient. Thus, for every index i between 0 and n and every index j between 0 and m consider prefixes $S[0 \ldots i]$ and $T[0 \ldots j]$ and let $\alpha_{i,j}$ be the value of an optimal alignment between strings $S[0 \ldots i]$ and $T[0 \ldots j]$.

3.2.3 Bellman Principle

If in an optimal alignment of strings $S[1 \ldots i]$ and $T[1 \ldots j]$ the last column of aligned symbols is deleted, one obtains an optimal alignment of strings $S[1 \ldots i-1]$ and $T[1 \ldots j-1]$ in case that $S(i)$ was aligned to $T(j)$, an optimal alignment of strings $S[1 \ldots i]$ and $T[1 \ldots j-1]$ in case that the spacing symbol was aligned to $T(j)$, and an optimal alignment of strings $S[1 \ldots i-1]$ and $T[1 \ldots j]$ in case that $S(i)$ was aligned to the spacing symbol.

3.2.4 Recursive Solution

Values $\alpha_{0,0}$, $\alpha_{i,0}$, $\alpha_{0,j}$, and $\alpha_{i,j}$, for $1 \leq i \leq n$ and $1 \leq j \leq m$, must be computed. We first treat the case of $\alpha_{i,j}$. Here we observe that at the right end of an optimal alignment of $S[0 \ldots i]$ with $T[0 \ldots j]$ either $S(i)$ is aligned with $T(j)$, or $S(i)$ is aligned with spacing symbol $-$, or spacing symbol $-$ is aligned with $T(j)$. We cannot predict which possibility gives the optimal scoring value, thus we must take the maximum of the three values (using here that Bellman's principle may be applied) $\sigma(S(i), T(j)) + \alpha_{i-1,j-1}$, $\sigma(S(i), -) + \alpha_{i-1,j}$, and $\sigma(-, T(j)) + \alpha_{i,j-1}$. For the computation of $\alpha_{i,0}$ we observe that there are no symbols available in $T[0 \ldots 0]$, thus there is only a single alignment of $S[0 \ldots i]$ with $T[0 \ldots 0]$ aligning spacing symbol to every symbol of $S[0 \ldots i]$ and having score equal to the sum of all values $\sigma(S(k), -)$, for $1 \leq k \leq i$. Correspondingly, $\alpha_{0,j}$ equals the sum of all values $\sigma(-, T(k))$, for $1 \leq k \leq j$. Finally we have to define basic value $\alpha_{0,0}$. Looking at the recursive formula above we observe that any occurrence of $\alpha_{0,0}$ must contribute value 0 to the overall score. This was also to be expected since the computation of $\alpha_{0,0}$ intends to optimally align empty string with empty string. Thus, we may summarize the recursive computation scheme (usually attributed to Needleman and Wunsch [56] under the name of "global pairwise alignment") as follows:

$$\alpha_{0,0} = 0$$

$$\alpha_{i,0} = \sum_{k=1}^{i} \sigma(S(k), -) = \alpha_{i-1,0} + \sigma(S(i), -)$$

$$\alpha_{0,j} = \sum_{k=1}^{j} \sigma(-, T(k)) = \alpha_{0,j-1} + \sigma(-, T(j)) \tag{3.1}$$

$$\alpha_{i,j} = \max \begin{cases} \alpha_{i,j-1} + \sigma(-, T(j)) \\ \alpha_{i-1,j} + \sigma(S(i), -) \\ \alpha_{i-1,j-1} + \sigma(S(i), T(j)). \end{cases}$$

3.2.5 Number of Different Subcalls and Overall Complexity

Given strings S and T of length n and m, respectively, there are $(n+1)(m+1)$ different sub-terms $\alpha_{i,j}$ to be computed, a number polynomial in n and m. Note that the computation of $\alpha_{i,0}$ requires a single call of an already computed value plus a single addition. The computation of $\alpha_{0,j}$ also requires a single call of an already computed value plus a single addition. Finally, the computation of $\alpha_{i,j}$ requires three calls of already computed values together with the computation of a maximum. In every case, a constant amount of work has to be done when computing a new value from already computed and stored values. Thus, the overall complexity of computing all sub-values is $O(nm)$. The final optimal score is found as value $\alpha_{n,m}$.

3.2.6 Tabular Organization of the Bottom-Up Computation

Whenever values $\alpha_{i,j}$ are to computed by hand we best organize the computation in a table as follows. The table has $n+1$ rows that are labelled with the spacing symbol and the symbols of string S, and $m+1$ columns that are labelled with the spacing symbol and the symbols of string T. First row and first column are initialized with values according to the first three of the equations (3.1) above. Inner field (i,j) receives its value using already computed values from three neighbouring fields according to the fourth of the equations (3.1) above. In addition, we use arrows indicating which of these three neighbouring fields are responsible to obtain the maximum value in the newly computed field; thus, there may be up to three arrows pointing to the new field. After finishing all computations, the arrows will help us to not only obtain the maximum value of an optimal alignment of S with T, but also an optimal alignment itself. For this, simply trace back arrows in an arbitrary manner from field (n,m) to field $(0,0)$; any diagonal step is interpreted as an alignment of two characters, whereas a vertical or horizontal step corresponds to the introduction of a spacing symbol in string T or S, respectively. We illustrate this with some examples. Consider strings $S =$AAAC and $T =$AGC with the scoring function that rewards matches by 1, penalizes mismatches by -1, and even stronger penalizes inserts and deletes by -2. The optimal score is read out as −1. Three different optimal alignments all having score −1 are obtained by three different traceback paths (Fig. 3.1).

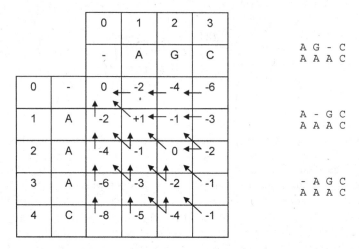

Fig. 3.1. Tabular computation of optimal alignment with traceback paths

Exercises

3.1. If instead mismatches are considerably stronger penalized than inserts and deletes by value −5, there result other optimal alignments than before. There are four different optimal alignments all having score −4. The reader may carry out the corresponding tabular computation.

$$
\begin{array}{cccc}
\text{A G -- -- C} & \text{A -- G -- C} & \text{A -- -- G C} & \text{-- -- A G C} \\
\text{A -- A A C} & \text{A A -- A C} & \text{A A A -- C} & \text{A A A -- C}
\end{array}
$$

3.2. Sometimes it is interesting to find an optimal alignment between prefixes and/or suffixes of one or both of strings S and T instead of the complete strings S and T. Show how all possible combinations of taking a prefix and/or suffix of one or both of the strings is treated computationally. Sometimes, the former computation may be used, whereas sometimes a minor modification must be built in (different initialization of fields with value 0).

3.3. Show how by a minor modification of the recursive formulas above an optimally scoring alignment between a substring of S and a substring of T can be found. Such an alignment is called an optimal *local alignment* (attributed to Smith and Waterman [70]).

3.3 Multiple Alignment

We show how the dynamic programming approach is generalized to more than two strings. We treat the example of three strings S_1, S_2, S_3 of lengths n_1, n_2, n_3. Define $\alpha_{i,j,k}$ to be the maximum obtainable sum-of-pairs score of an alignment of prefixes $S_1[1 \ldots i]$, $S_2[1 \ldots j]$, $S_3[1 \ldots k]$. We concentrate on the case $i > 0$, $j > 0$, $k > 0$, and leave the correct definitions for base cases with all or two or one of parameters i, j, k being 0 to the reader. As before, we distinguish what may happen at the right end of an optimal alignment. Obviously, there are 7 ways to introduce up to two spacing symbols. The formula thus looks as follows:

$$
\alpha_{i,j,k} = \max \begin{cases}
\alpha_{i-1,j-1,k-1} + \sigma(S_1(i), S_2(j)) + \sigma(S_1(i), S_3(k)) + \sigma(S_2(j), S_3(k)) \\
\alpha_{i,j-1,k-1} + \sigma(-, S_2(j)) + \sigma(-, S_3(k)) + \sigma(S_2(j), S_3(k)) \\
\alpha_{i-1,j,k-1} + \sigma(S_1(i), -) + \sigma(S_1(i), S_3(k)) + \sigma(-, S_3(k)) \\
\alpha_{i-1,j-1,k} + \sigma(S_1(i), S_2(j)) + \sigma(S_1(i), -) + \sigma(S_2(j), -) \\
\alpha_{i,j,k-1} + \sigma(-, S_3(k)) + \sigma(-, S_3(k)) \\
\alpha_{i,j-1,k} + \sigma(-, S_2(j)) + \sigma(S_2(j), -) \\
\alpha_{i-1,j,k} + \sigma(S_1(i), -) + \sigma(S_1(i), -).
\end{cases}
$$

$$(3.2)$$

In the general case of k strings of length $O(n)$ each, there are n^k terms to be computed, with max terms consisting of $2^k - 1$ sub-terms. This leads to overall complexity of $O(n^k 2^k)$. This makes the approach impractical even for small values of k and n.

3.4 Affine Gap Alignment

3.4.1 Problem Restated

In an alignment like the following it might be desirable to penalize the insertion or deletion of blocks of length k not as strongly as k independent insertions or deletions.

$$
\begin{aligned}
&A\,A\,C - - - A\,A\,T\,T\,C\,C\,G\,A\,C\,T\,A\,C \\
&A\,C\,T\,A\,C\,C\,T - - - - - - C\,G\,C - -
\end{aligned}
$$

We divide an alignment in blocks as follows: *a-blocks* consist of a single pair of characters both different from the spacing symbol; *b-blocks* are segments of maximum length consisting of only spacing symbols in the first string; *c-blocks* are segments of maximum length consisting of only spacing symbols in the second string. In the example above we observe in consecution three a-blocks, a single b-block of length 3, a single a-block, a single c-block of length 6, three a-blocks, and a single c-block of length 2. We make the convention that b-blocks and c-blocks must always be separated by at least one a-block. Any a-block aligning characters x and y is scored as usual by $\sigma(x, y)$, whereas any b-block or c-block of length k is scored as a whole by $-f(k)$, where $f(k)$ is a function that usually is sublinear, that is $f(k) \leq k$. Examples of gap scoring functions are $f(k) = \log k$, or any linear function $f(k) = ck$, or any affine function $f(k) = e(k-1) + d$. We illustrate the modifications necessary to deal with gaps for the case of affine gap scoring, though the approach also works for an arbitrary function $f(k)$.

AFFINE GAP ALIGNMENT
Given scoring function $\sigma(x, y)$ for characters, affine gap penalizing function for b-blocks and c-blocks of length k, $f(k) = -e(k-1) - d$, and strings S, T of length n and m, respectively, compute the score of an optimal gap alignment $\sigma_{\mathrm{opt}}(S, T)$.

3.4.2 Parameterization and Conditioning

Parameterization is obvious; for every i and j with $0 \leq i \leq n$ and $0 \leq j \leq m$ optimally align prefixes $S[1 \ldots i]$ and $T[1 \ldots j]$. The problem comes with the way gaps are scored: if an alignment ends, for example, with a pair $(x, -)$ it may not be locally scored without taking into consideration what is left of it. In case that left of it a further pair $(y, -)$ is found, rightmost pair $(x, -)$ contributes $-e$ to the affine score of the whole b-block. In case that left of it an a-block (y, z) is found, rightmost pair $(x, -)$ contributes $-d$ to the affine score of the whole b-block. Information on what is left of it may be controlled by simultaneously optimizing three functions instead of a single one. These are described as follows:

- $a_{i,j}$ = maximum value of a gap alignment of $S[1 \ldots i]$ and $T[1 \ldots j]$ ending with an a-block
- $b_{i,j}$ = maximum value of a gap alignment of $S[1 \ldots i]$ and $T[1 \ldots j]$ ending with a b-block
- $c_{i,j}$ = maximum value of a gap alignment of $S[1 \ldots i]$ and $T[1 \ldots j]$ ending with a c-block

After having computed all of these values the problem is solved by taking as optimal gap alignment value $\sigma_{\mathrm{opt}}(S, T) = \max\{a_{n,m}, b_{n,m}, c_{n,m}\}$.

3.4.3 Bellman Principle

... is obviously fulfilled.

3.4.4 Recursive Solution

In the following recursive formulas we have $i > 1$ and $j > 1$. Non-admissible parameter combinations are ruled out by using $-\infty$ as value.

$$a_{1,1} = \sigma(S(1), T(1))$$
$$a_{i,1} = \sigma(S(i), T(1)) + c_{i-1,0}$$
$$a_{1,j} = \sigma(S(1), T(j)) + b_{0,j-1}$$
$$a_{i,j} = \sigma(S(i), T(j)) + \max\{a_{i-1,j-1}, b_{i-1,j-1}, c_{i-1,j-1}\}$$

$$b_{0,1} = -d$$
$$b_{0,j} = -e - b_{0,j-1}$$
$$b_{1,1} = -\infty \ (\text{impossible case})$$
$$b_{i,1} = -\infty \ (\text{impossible case})$$
$$b_{1,j} = \max\{-e + b_{1,j-1}, -d + a_{1,j-1}\} \qquad (3.3)$$
$$b_{i,j} = \max\{-e + b_{i,j-1}, -d + a_{i,j-1}\}$$

$$c_{1,0} = -d$$
$$c_{i,0} = -e - c_{i-1,0}$$
$$c_{1,1} = -\infty \ (\text{impossible case})$$
$$c_{1,j} = -\infty \ (\text{impossible case})$$
$$c_{i,1} = \max\{-e + c_{i-1,1}, -d + a_{i-1,1}\}$$
$$c_{i,j} = \max\{-e + c_{i-1,j}, -d + a_{i-1,j}\}$$

3.4.5 Number of Different Subcalls and Overall Complexity

Counting the steps that are required to compute all values $a_{i,j}$, $b_{i,j}$, and $c_{i,j}$, we see that for the calculation of a further value we require at most three already computed values. Thus, overall complexity to compute all required values is again $O(nm)$.

Exercises

3.4. What has to be modified if we allow b-blocks and c-blocks to occur side by side?

3.5. Modify the equations above for the case of an arbitrary gap penalizing function $f(k)$ that depends on the length k of b-blocks and c-blocks. Show that overall complexity increases to $O(n^2m + nm^2)$.

3.5 Exon Assembly

3.5.1 Problem Restated

From a set of candidate exons a selection of non-overlapping ones is to be found whose concatenation in ascending order best fits to a proposed target string.

EXON ASSEMBLY

Given scoring function σ with $\sigma(x, -) \leq 0$ and $\sigma(-, x) \leq 0$ for all characters x (as it is usually the case for standard scoring functions), genome string $G = G[1 \ldots n]$, list of candidate exons, i.e. substrings $E_1, \ldots, E_b$ of G, and target string $T = T[1 \ldots m]$, compute the best possible optimal alignment score for any concatenation Γ^* of a chain Γ of candidate exons with T:

$$\max_{\text{chains } \Gamma} \sigma_{\text{opt}}(\Gamma^*, T).$$

The following notions are used in the sections below. For a candidate exon $E = G[i \ldots j]$ define $\text{first}(E) = i$ and $\text{last}(E) = j$. For candidate exons E and F let $E < F$ stand for $\text{last}(E) < \text{first}(F)$.

3.5.2 Parameterization and Conditioning

As usual, parameterization is done via truncation of strings G and T. Concerning target string T, we explicitly truncate it at some position j with $0 \leq j \leq m$. Concerning genome string G, truncation is done indirectly by fixing some exon candidate E_k as the last one to be used in any chain of candidate exons, and truncating E_k at some position i with $\text{first}(E_k) \leq i \leq \text{last}(E_k)$. Note that truncation of T may lead to the empty string (in case $j = 0$), whereas truncation of E_k always gives a non-empty string (containing at least base $G(i)$). For a chain Γ of candidate exons $F_1 < F_2 < \ldots < F_p$ and index i with $\text{first}(F_p) \leq i \leq \text{last}(F_p)$, we use the suggestive notation $\Gamma^*[\ldots i]$ to denote the concatenation $F_1 F_2 \ldots F_{p-1} G[\text{first}(F_p) \ldots i]$, with the last candidate exon truncated at absolute position i.

Now the parameterized and conditioned problem can be stated as follows: for all combinations of parameters j, k, i with $0 \leq j \leq m$, $1 \leq k \leq b$, and $\text{first}(E_k) \leq i \leq \text{last}(E_k)$ compute the following value chain(j, k, i):

$$\text{chain}(j, k, i) = \max_{\Gamma \text{ ending with } E_k} \sigma_{\text{opt}}(\Gamma^*[\ldots i], T[1 \ldots j]). \qquad (3.4)$$

After computation of all these values, the original problem is then solved by returning

$$\max_{1 \leq k \leq b} \text{chain}(m, k, \text{last}(B_k)). \qquad (3.5)$$

3.5.3 Bellman Principle

... is obviously fulfilled.

3.5.4 Recursive Solution

The solution presented here is taken from [64]. For the computation of terms chain(j, k, i) for any admissible combination of parameters j, k, i we distinguish several cases:

Case 1. $j = 0$
Case 2. $j > 0$, there is no index r with $E_r < E_k$
Case 3. $j > 0$, first(E_k) $< i$, there is at least one index r with $E_r < E_k$
Case 4. $j > 0$, first(E_k) $= i$, there is at least one index r with $E_r < E_k$

In case 1, we have to optimally align a suitable string $\Gamma^*[\ldots i]$ with the empty string $T[1\ldots 0]$. Since inserts and deletes are scored zero or negative, the optimal alignment is achieved by using no other candidate exons than the prescribed last E_k. We obtain:

$$\text{chain}(0, k, i) = \sum_{p=\text{first}(B_k)}^{i} \sigma(G(p), -). \tag{3.6}$$

In case 2, there are no further candidate exons left of E_k that can be used in $\Gamma^*[\ldots i]$. Thus we obtain:

$$\text{chain}(j, k, i) = \sigma_{\text{opt}}(G[\text{first}(E_k)\ldots i], T[1\ldots j]). \tag{3.7}$$

In case 3, we look at what may happen at the right end of an optimal alignment of $\Gamma^*[\ldots i]$ and $T[1\ldots j]$. As usual, either $G(i)$ is aligned with $T(j)$, or $G(i)$ is aligned with spacing symbol $-$, or spacing symbol $-$ is aligned with $T(j)$. In any case, in the recursive step we do not leave E_k as last used exon candidate. Thus we obtain:

$$\text{chain}(j, k, i) = \max \begin{cases} \text{chain}(j-1, k, i-1) + \sigma(G(i), T(j)) \\ \text{chain}(j, k, i-1) + \sigma(G(i), -) \\ \text{chain}(j-1, k, i) + \sigma(-, T(j)). \end{cases} \tag{3.8}$$

In case 4, after aligning $G(i)$ with $T(j)$ or $G(i)$ with spacing symbol $-$, we have consumed the last available symbol of E_k. Every candidate exon E_q with $E_q < E_k$ may be the segment used in Γ left of E_k; alternatively E_k was the only segment used in Γ. Thus we obtain:

$$\text{chain}(j, k, i) = \max \begin{cases} \sigma(G(i), T(j)) + \max\limits_{q \text{ with } \text{last}(B_q)<i} \text{chain}(j-1, q, \text{last}(B_q)) \\ \sigma(G(i), -) + \max\limits_{q \text{ with } \text{last}(B_q)<i} \text{chain}(j, q, \text{last}(B_q)) \\ \sigma(G(i), T(j)) + \sum_{p<j} \sigma(-, T(p)) \\ \sigma(G(i), -) + \sum_{p\leq j} \sigma(-, T(p)) \\ \sigma(-, T(j)) + \text{chain}(j-1, k, i). \end{cases}$$

$$\tag{3.9}$$

3.5.5 Number of Different Subcalls and Overall Complexity

There are $O(bnm)$ terms chain(j, k, i) to be computed. Every computation of a new such value using stored prior values requires the following efforts. To find out which of the four cases above applies a run through the list of candidate exons is required. This costs $O(b)$ steps. In case 1, there are $O(n)$ additions to be done. In case 2, $O(nm)$ steps are required to compute an optimal alignment. Case 3 is cheapest, only $O(1)$ steps lead to the desired result. Case 4 requires $O(b)$ already computed values to be used in the computation of the maxima, and $O(m)$ additions to be done in the computations of the sums. Altogether, we have a complexity of $O(bnm(b + n + nm + b + m)) = O(bn^2m^2 + b^2nm)$.

3.5.6 Preprocessing of Simple Terms

There is some waste of time in the procedure described above. This concerns a couple of terms that are computed in the calls of chain(j, k, i) in a multiple manner. Doing the computation of these terms in advance saves execution time. The terms are the following. If we compute for every $i \leq n$ the list of all exon candidates E_q with last$(E_q) < i$, and also the list of all exon candidates E_q with last$(E_q) = i$, the distinction of cases can be done in $O(1)$ time. The computation of these lists in advance requires $O(bn)$ steps. Sums occurring in case 1 can be computed in advance with $O(bn)$ additions. For a fixed value first(E_k), optimal alignment values for all combinations of strings $E_k[\text{first}(E_k) \ldots i]$ and $T[1 \ldots j]$ occurring in case 2 can be computed in $O(nm)$ steps using the tabular computation of an optimal alignment of $E_k[\text{first}(E_k) \ldots \text{last}(E_k)]$ and $T[1 \ldots m]$. Thus, all values required for case 2 can be computed in $O(bnm)$ steps. Finally, the sums occurring in case 4 are all computed in $O(m)$ steps. What cannot be treated this way are the maxima in case 4 since these by themselves contain further recursive calls of function chain, i.e. are not simple terms. Thus, $O(b)$ steps are required to compute chain(j, k, i) via case 4 from already computed values. We end with an overall complexity of $O(bn + bn + bnm + m + nmb^2)$. This simplifies to $O(b^2nm)$ and improves the former estimation.

3.5.7 Parallel Computation of Recursive Terms

To also get rid of the remaining extra factor b in the estimation above requires the separate computation of just those terms in case 4 which caused this extra factor. Let us abbreviate them as follows. For every combination of parameters i and j such that there is at least one exon candidate E_q with last$(E_q) < i$ define

$$\mu(i, j) = \max_{q \text{ with last}(E_q) < i} \text{chain}(j, q, \text{last}(E_q)). \tag{3.10}$$

Using this abbreviation, case 4 can now be restated as:

$$\text{chain}(j,k,i) = \max \begin{cases} \sigma(G(i), T(j)) + \mu(i, j-1) \\ \sigma(G(i), -) + \mu(i, j) \\ \sigma(G(i), T(j)) + \sum_{p<j} \sigma(-, T(p)) \\ \sigma(G(i), -) + \sum_{p \leq j} \sigma(-, T(p)) \\ \sigma(-, T(j)) + \text{chain}(j-1, k, i). \end{cases} \quad (3.11)$$

Now we see that a constant number of calls of stored values is required instead of the $O(b)$ amount before. But of course we must show how to recursively compute also all fresh terms $\mu(i, j)$; simply introducing an abbreviation alone is not a clever trick for saving execution time. The recursive computation of $\mu(i, j)$ makes the following distinctions:

Case 1. There is no E_q with $\text{last}(E_q) = i - 1$.

Case 2. There is at least one E_q with $\text{last}(E_q) = i - 1$, and at least one E_p with $\text{last}(E_p) < i - 1$.

Case 3. There is at least one E_q with $\text{last}(E_q) = i - 1$, but no E_p with $\text{last}(E_p) < i - 1$.

In the first case we obtain:

$$\mu(i, j) = \mu(i - 1, j). \quad (3.12)$$

In the second case we obtain:

$$\mu(i, j) = \max \begin{cases} \mu(i - 1, j) \\ \max\limits_{q \text{ with last}(E_q)=i-1} \text{chain}(j, q, i - 1). \end{cases} \quad (3.13)$$

Note that the first term in the outer max covers all E_q with $\text{last}(E_q) < i - 1$, whereas the inner max term covers all E_q with $\text{last}(E_q) = i - 1$. In the third case we obtain:

$$\mu(i, j) = \max\limits_{q \text{ with last}(E_q)=i-1} \text{chain}(j, q, i - 1). \quad (3.14)$$

The computation of all values $\mu(i, j)$ obviously requires $O(bnm)$ steps with factor b resulting from the max terms in cases 2 and 3. Summarizing the complexity of computation of all values that are stored in advance is $O(bn + m + bnm + bn) = O(bnm)$, the computation of all values $\text{chain}(i, j, k)$ and $\mu(i, j)$ requires $O(bnm) + O(bnm) + O(bnm) = O(bnm)$ steps. This is optimal since there are bnm terms to be computed.

3.6 RNA Structure Prediction

3.6.1 Problem Restated

We want to start with the basic problem of computing the pseudoknot-free structure with maximum number $\beta(1, n)$ of base pairs for an RNA string $S = S[1 \ldots n]$. The more challenging task of computing the pseudoknot-free structure with minimum free energy will be discussed in Sect. 3.6.5.

3.6.2 Parameterization and Conditioning

Let $\beta(i,j)$ be the maximum number of base pairs in a pseudoknot-free folding of subsequence $S[i \ldots j]$. As the following recursive solution shows, consideration of all prefixes of S like in alignment or exon assembly would not be sufficient here. We want to calculate the best structure for subsequence $S[i \ldots j]$ from the previously calculated best structure for smaller subsequences.

3.6.3 Recursive Solution and Bellman Principle

The computation of $\beta(i,j)$ distinguishes the following cases.

1. Add an unpaired base i to the best structure for smaller subsequence $S[i+1 \ldots j]$.

2. Add an unpaired base j to the best structure for smaller subsequence $S[i \ldots j-1]$.

3. Add a base pair (i,j) with score $\delta(i,j)$ to the best structure for smaller subsequence $S[i+1 \ldots j-1]$.

4. Combine two best structures for smaller subsequences $S[i \ldots k]$ and $S[k+1 \ldots j]$.

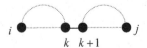

In case 1, $\beta(i,j)$ obviously equals $\beta(i+1,j)$. In case 2, $\beta(i,j)$ obviously equals $\beta(i,j-1)$. In case 3, binding of base i to base j contributes a score $\delta(i,j)$ to $\beta(i,j)$. Note that this score can either be defined as a constant or as a free energy value reflecting how strong the bond between this specific base pair is. In case 4, we construct a so-called *bifurcation* with two best structures on each side. Summarizing all cases, we obtain the following equations called the *Nussinov algorithm* [60], for $1 \leq i \leq n$ and $i < j \leq n$:

$$\beta(i,i) = 0 \quad \text{for } i = 1,\ldots,n$$
$$\beta(i,i-1) = 0 \quad \text{for } i = 2,\ldots,n$$
$$\beta(i,j) = \max \begin{cases} \beta(i+1,j) \\ \beta(i,j-1) \\ \beta(i+1,j-1) + \delta(i,j) \\ \max_{i<k<j}\{\beta(i,k) + \beta(k+1,j)\}. \end{cases} \quad (3.15)$$

Note that the additive term $\delta(i,j)$ can rule out pairings that are not admissible. For example, set

$$\delta(i,j) = \begin{cases} 1 & \text{if } (i,j) = (A, U) \text{ or } (C, G) \\ 0 & \text{else.} \end{cases}$$

In practice, a matrix will be filled along the diagonals and the solution can be recovered through a traceback step. Figure 3.2 visualizes how a matrix entry is computed recursively. Note that only the upper (or lower) half of the matrix needs to be filled. Therefore, after initialization the recursion runs from smaller to longer subsequences as follows:

```
for l = 1 to n do
   for i = 1 to n + 1 - l do
      j = i + l
      compute β(i, j)
   end for
end for
```

3.6.4 Number of Different Subcalls and Overall Complexity

There are $O(n^2)$ terms to be computed, each requiring calling of $O(n)$ already computed terms for the case of bifurcation. Thus overall complexity is $O(n^3)$ time and $O(n^2)$ space.

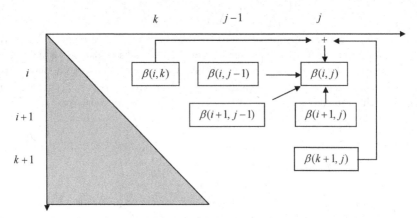

Fig. 3.2. Calculation of a matrix entry in the *Nussinov* algorithm

3.6.5 Variation: Free Energy Minimization

The base pair maximization approach neglects two main factors which in nature strongly drive RNA folding. First, loop sizes (entropic terms) are not taken into account. This is critical, as long unstructured loop regions destabilize and are therefore unlikely to form in a structure which aims for the state of minimum free energy. Second, *helical stacking* is ignored in the Nussinov algorithm. The enthalpic term which stacked base pairs (i, j) and $(i+1, j-1)$ contribute to stabilize a structure varies considerably depending on which bases are involved. For example, experimental work showed that helical stacking of pairs (G,U) and (U,G) has free energy of around -0.2 kcal/mol, whereas helical stacking of (U,G) and (G,U) is assigned free energy of around -1.5 kcal/mol. Note that helical stacking of pairs (i, j) and $(i + 1, j - 1)$ is not a symmetric relation.

In order to take into account global folding motifs such as stems, hairpin loops, bulges, internal loops, and multiloops (Sects. 1.6 and 2.9) we need to describe them in a formal notion. Thinking of loop regions, the most promising way to identify a structure element is by its *closing base pair*, i.e. the bond with largest distance. Let R be an RNA structure over a sequence S. We say a base k is *accessible* from a closing base pair $(i, j) \in R$ if there are no other base pairs $(i', j') \in R$ such that $i < i' < k < j' < j$. Informally stated, this means that if we want to reach an accessible base from the closing base pair, there is no other base pair in the way. Similarly, a base pair $(k, l) \in R$ is *accessible* from a closing base pair (i, j) if both k and l are. We call (k, l) an *interior base pair*. Now, let us define a structure element by its closing bond and interior base pairs as follows (Fig. 3.3):

- A loop with closing base pair $(i, j) \in R$ and one interior base pair $(i + 1, j - 1) \in R$ is called a *stem*.

- A loop with closing base pair $(i, j) \in R$ and no interior base pairs is called a *hairpin loop*.
- A loop with a closing base pair $(i, j) \in R$ and one interior base pair $(i', j') \in R$ with $(i' - i) + (j - j') > 2$ is called an *internal loop*.
- An internal loop is called a *bulge loop*, if $j' = j - 1$ or $i' = i + 1$.
- A loop with interior base pairs $(i_1, j_1) \ldots (i_k, i_k) \in R$ together with a closing base pair $(i, j) \in R$ is called a *k-multiloop*.

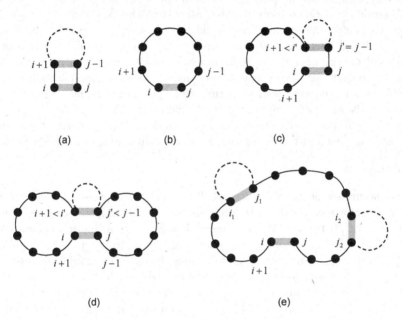

Fig. 3.3. Structure elements defined by their closing bond (i, j): **(a)** stem; **(b)** hairpin loop; **(c)** bulge; **(d)** internal loop; **(e)** multiloop (*hydrogen bonding is indicated by gray area*)

For each of the structure elements, an energy function is defined which depends on the closing bond (i, j) and interior base pairs:

- $eS(i, j, i + 1, j - 1)$ is the energy of a *stem* closed by (i, j).
- $eH(i, j)$ is the energy of a *hairpin loop* closed by (i, j).
- $eL(i, j, i', j')$ is the energy of an *internal loop* or *bulge loop* closed by (i, j).
- $eM(i, i_1, j_1, \ldots, i_k, j_k, j)$ is the energy of a *k-multiloop* closed by (i, j) with interior base pairs $(i_1, j_1) \ldots (i_k, j_k)$. To make the energy computation tractable and avoid exponential runtime, the following *k-multiloop* energy simplification is commonly made:

$$eM(i, i_1, j_1, \ldots, i_k, j_k, j) = a + bk + ck' \qquad (3.16)$$

with $a, b, c = $ constants and $k' = $ number of unpaired bases in the loop.

We fix the notation that every secondary structure element identified by its closing bond (i, j) contributes an energy value $E_{i,j}^R$ (calculated from the underlying energy function) to the overall free energy $E(R)$. The free energy of an RNA structure R can now be calculated as the additive sum:

$$E(R) = \sum_{(i,j) \in R} E_{i,j}^R. \qquad (3.17)$$

It is now clear that computing the minimum free energy structure for an RNA sequence is much more sophisticated than the *Nussinov* approach we discussed before. We simply maximized the number of base pairs (and possibly produced long unstructured loop regions). Here, we take the sum over individual energy contributions from structure elements such as stems and loops. Note that this free energy minimization approach fixes a certain *energy model*, namely the sum over structure element energy values. Therefore, it remains only an approximation of RNA folding. It allows an elegant framework for computational methods, however there are surely more complicated folding processes and factors hidden somewhere in the complex RNA world.

Zuker algorithm

The structure elements and corresponding energy model described above are the scaffold for a clever dynamic programming algorithm, the so-called *Zuker algorithm* [81]. It takes an RNA sequence $S = S[1 \dots n]$ as input and computes the structure R with minimum free energy according to the underlying energy model. In the algorithm, three functions must be optimized simultaneously. In practice, these correspond to matrices which we need to fill. First, we define $\epsilon(i)$ to hold the minimum free energy of a structure on subsequence $S[1 \dots i]$. Second, we denote $\pi(i, j)$ to hold the minimum free energy of a structure on subsequence $S[i \dots j]$ *with i and j paired*. Third, another conditioned function is needed to account for multiloops. We define $\mu(i, j)$ to hold the minimum free energy of a structure on subsequence $S[i \dots j]$ that is *part of a multiloop*.

We start with the description of $\pi(i, j)$, which demands that bases i and j form a closing base pair for the structure element on $S[i \dots j]$ with minimum free energy. The following four cases have to be distinguished:

1. Base pair (i, j) closes a *stem*. We need to add the stacking energy $eS(i, j, i + 1, j - 1)$ and minimum free energy of a structure on smaller subsequence $S[i + 1 \dots j - 1]$ *closed by base pair $(i + 1, j - 1)$*.

2. Base pair (i, j) closes a *hairpin loop*. We need to add the hairpin loop energy $eH(i, j)$.

3. Base pair (i, j) closes a *bulge* or *internal loop*. We need to add the loop energy $eL(i, j, i', j')$ and minimum free energy of a structure on smaller subsequence $S[i' \ldots j']$ closed by base pair (i', j').

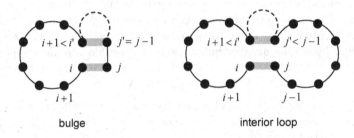

bulge   interior loop

4. Base pair (i, j) closes a *multiloop*. We need to decompose the multiloop into two smaller subsequences $S[i+1 \ldots k-1]$ and $S[k \ldots j-1]$ (described in more detail below) and add the offset penalty a.

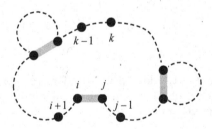

If bases i and j cannot form an admissible base pair, we set $\pi(i,j) = \infty$. The four cases lead to the following recurrence for all i, j with $1 \leq i < j \leq n$:

$$\pi(i,j) = \min \{E(R) |\ R \text{ structure for } S[i \ldots j] \land (i,j) \in R\}$$

$$= \min \begin{cases} eS(i,j,i+1,j-1) + \pi(i+1,j-1) \\ eH(i,j) \\ \min_{\substack{i<i'<j'<i \\ i'-i+j-j'>2}} \{eL(i,j,i',j') + \pi(i',j')\} \\ \min_{i+1<k\leq j-1} \{\mu(i+1,k-1) + \mu(k,j-1) + a\}. \end{cases} \qquad (3.18)$$

We initialize $\pi(i, i-1) = \pi(i,i) = \infty$, as these base pairs cannot form in a structure. Now, let us take a further look at case 4, the multiloop calculation. We already treated the closing base pair (i,j) in the $\pi(i,j)$ calculation. Searching for the multiloop structure with minimum free energy amongst all possible interior base pairs would lead to exponential runtime. Therefore, we use a simple trick to make the computation feasible. We recursively cut the two parts of a multiloop down to the interior base pairs by moving one base at a time or by adding another bifurcation. We also have to include penalties according to the underlying energy simplification (3.16). The following four cases have to be distinguished for $\mu(i,j)$:

1. We move to base $i + 1$, add a penalty c for an unpaired base, and the minimum free energy of a structure on smaller subsequence $S[i+1 \ldots j]$ that is part of a multiloop.

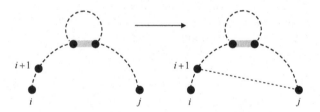

2. We move to base $j - 1$, add a penalty c for an unpaired base, and the minimum free energy of a structure on smaller subsequence $S[i \ldots j-1]$ that is part of a multiloop.

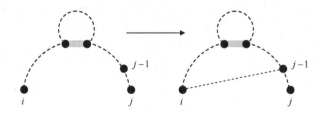

3. We perform another multiloop bifurcation at base k and add the minimum free energy of two structures on smaller subsequences $S[i \ldots k-1]$ and $S[k \ldots j]$ that are part of a multiloop.

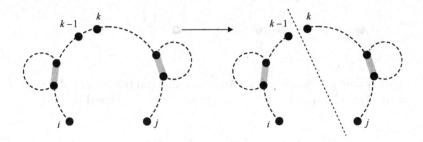

4. We discover an interior base pair (i, j) and obtain the minimum free energy of a structure on subsequence $S[i \ldots j]$ closed by (i, j).

The four cases lead to the following recurrence for all i, j with $1 \le i < j \le n$:

$$\mu(i, j) = \min \{E(R)| \ R \text{ structure for } S[i \ldots j] \text{ that is part of a multiloop}\}$$

$$= \min \begin{cases} \mu(i+1, j) + c \\ \mu(i, j-1) + c \\ \min_{i<k\le j} \{\mu(i, k-1) + \mu(k, j)\} \\ \pi(i, j) + b. \end{cases}$$

$$(3.19)$$

In order to ensure that we really produce at least two interior base pairs in a multiloop, we must initialize as follows: $\mu(i, i) = \infty$.

So far we are able to produce structure elements with a spanning closing base pair. However, there is no way to arrange structures in a consecutive fashion. Therefore, a third function $\epsilon(i)$ is introduced in the *Zuker* algorithm, which holds the minimum free energy of a structure on subsequence $S[1 \ldots i]$. The computation of $\epsilon(i)$ distinguishes the following two cases:

1. Add an unpaired base i to the best structure for smaller subsequence $S[1 \ldots i-1]$.

2. Base i is paired to some base k. Add the minimum free energy of structures on smaller subsequences $S[1 \ldots k-1]$ and $S[k \ldots i]$ closed by (k, i).

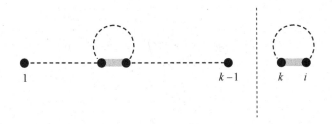

Taking together both cases, we obtain the following formula for $\epsilon(i)$ with $1 \leq i \leq n$:

$$\epsilon(i) = \min \{E(R) | \ R \text{ structure for } S[1 \ldots i]\}$$

$$= \min \begin{cases} \epsilon(i-1) \\ \min_{1 \leq k \leq i} \{\epsilon(k-1) + \pi(k, i)\}. \end{cases} \tag{3.20}$$

We have to initialize as follows: $\epsilon(0) = 0$.

Now we have completely described the *Zuker* algorithm for computing an RNA structure with minimum free energy in the three equations (3.17), (3.18), and (3.19). In practice, three matrices $\epsilon(i)$, $\pi(i, j)$, and $\mu(i, j)$ are filled using the principle of dynamic programming for all i, j with $1 \leq i < j \leq n$. When all entries are computed, $\epsilon(n)$ contains the minimum free energy for an RNA sequence $S = S[1 \ldots n]$ and the corresponding best structure can be recovered by a traceback path.

Let us analyze the time and space requirements. Obviously, the entries for matrix $\epsilon(i)$ can be computed in $O(n^2)$ time and $O(n)$ space. It takes $O(n^3)$ time and $O(n^2)$ space to fill matrix $\mu(i, j)$. The most critical is matrix $\pi(i, j)$. Take a look at the case for bulges and internal loops. The minimum calculation requires $O(n^4)$ time and $O(n^2)$ space, as we have to minimize over two positions i' and j'. However, with certain internal loop length restrictions, the *Zuker* algorithm requires $O(n^3)$ time and $O(n^2)$ space to find the structure with minimum free energy.

It is easy to see that the *Zuker* algorithm excludes pseudoknots. The structure with lowest free energy for the subsequence $S[i \ldots j]$ is only allowed to contain non-crossing interactions within this interval. Otherwise, the dynamic programming principle is violated and efficient computation becomes impossible. Pseudoknot interaction happens in a crossing fashion and therefore dynamic programming algorithms for RNA secondary structure prediction neglect them.

3.7 Viterbi Algorithm

3.7.1 Problem Restated

Let $M = (Q, \Sigma, q_0, T, E)$ be a Hidden Markov model with set of states Q, alphabet Σ of emitted characters, start state q_0, state transition probabilities $T(p,q)$, and character emission probabilities $E(q,x)$. Given an observed sequence $S \in \Sigma^n$, the Viterbi algorithm computes the most probable state sequence $W \in Q^n$ that might have emitted string S.

> **HMM DECODING**
> Given string $S \in \Sigma^n$, compute $W \in Q^n$ with maximal value $P(W, S)$.

3.7.2 Parameterization and Conditioning

For each $q \in Q$ and $i = 1, \ldots, n$ define:

$$\mu_i(q) = \max_{W \in Q^{i-1}} P(Wq, S[1 \ldots i]). \tag{3.21}$$

Here, conditioning is on the last used state q in a state sequence that emitted prefix $S[1 \ldots i]$.

3.7.3 Bellman Principle

... is obviously fulfilled.

3.7.4 Recursive Computation

$$\begin{aligned}
\mu_1(q) &= T(q_0, q) E(q, S(1)) \\
\mu_{i+1}(q) &= E(q, S(i+1)) \max_{r \in Q} \{\mu_i(r) T(r, q)\}.
\end{aligned} \tag{3.22}$$

Having computed these values the original task is solved by:

$$\mu_S = \max_{q \in Q} \mu_n(q). \tag{3.23}$$

3.7.5 Counting and Overall Complexity

Obviously, $O(n|Q|^2)$ steps are required to compute all these values $\mu_i(q)$.

3.8 Baum-Welch Algorithm

3.8.1 Problem Motivation

Next we discuss an often used way to fit parameters of a Hidden Markov model M to observations consisting of strings $S_1, \ldots, S_m$. The Viterbi algorithm computes the most likely state strings $W_1, \ldots, W_m$. Afterwards, these together with $S_1, \ldots, S_m$ are used to count transition and emission frequencies which are used as fresh values for an updated model M^*. Though being intuitive, the disadvantage of this approach is that state strings are used that maximize joint probability, and working with maxima is notoriously difficult in probability theory. It is much better to work with expectations. In our context this means that we use all state sequences to update transition and emission probabilities, however each state sequence is weighted with its probability of occurrence. Thus, plausible sequences are preferred via their probability of occurrence. For example, we thus have to compute the expected transition frequency between any two states, the expectation taken over all state sequences. If this is done, one can indeed formally prove that for the updated model M^* likelihood of emitting $S_1, \ldots, S_m$ is improved compared to model M. This is an example of a more fundamental algorithm, called *expectation maximization* (EM) algorithm.

3.8.2 A Couple of Important 'Variables'

In probability theory, a function defined on a probability space usually is called a 'variable'. Given a Hidden Markov model M, the following variables play an important role. For a fixed emitted string $S \in \Sigma^n, 1 \le i \le n$ and $1 \le j < n$ we define them as follows:

- *Forward variable* $\alpha_i(q)$ as the probability that M generates a state sequence with i^{th} state q and emits a symbol sequence with prefix $S[1 \ldots i]$.
- *Backward variable* $\beta_i(q)$ as the probability that M emits a symbol sequence with suffix $S[i+1 \ldots n]$ provided it generated a state sequence with i^{th} state q.
- *Transition variable* $\eta_i(q, r)$ as the probability that M generates a state sequence with j^{th} state q and $(j+1)^{\text{th}}$ state r provided it emitted symbol sequence S.
- *State variable* $\gamma_i(q)$ as the probability that M generates a state sequence with i^{th} state q provided it emitted symbol sequence S.

3.8.3 Computing Forward Variables

The forward variables are computed in $O(|S||Q|^2)$ steps as follows:

$$\alpha_1(q) = p(q)E(q, S(1))$$
$$\alpha_{i+1}(q) = E(q, S(i+1)) \sum_{r \in Q} \alpha_i(r)T(r, q). \tag{3.24}$$

As an application of forward variables, we see that emission probabilities $P(S)$ of string S of length n (original definition required an exponential summation) can be computed efficiently:

$$P(S) = \sum_{q \in Q} \alpha_n(q). \tag{3.25}$$

3.8.4 Computing Backward Variables

The backward variables are computed similarly in $O(|S||Q|^2)$ steps as follows:

$$\beta_n(q) = 1$$
$$\beta_{i-1}(q) = \sum_{r \in Q} T(q, r)E(r, S(i))\beta_i(r). \tag{3.26}$$

Exercise

3.6. Express $P(S)$ using backward variable instead of forward variable as in Sect. 3.8.3.

3.8.5 Computing Transition Variables

Now we can obtain the transition variables on basis of the following equation. Its left-hand side is the probability of observing string S and having states q and r at positions j and $j + 1$, whereas the right-hand side decomposes this event into observing prefix $S[1 \ldots j]$ and having state q at position j, then switching to state r at position $j + 1$ and emitting $S(j + 1)$, and finally observing suffix $S[j + 1 \ldots n]$ after starting with state r at position $j + 1$.

$$\eta_j(q, r)P(S) = \alpha_j(q)T(q, r)E(r, S(j+1))\beta_{j+1}(r) \tag{3.27}$$

3.8.6 Computing State Variables

$$\gamma_1(q) = P(q)$$
$$\gamma_{i+1}(q) = \sum_{r \in Q} \eta_i(r, q) \tag{3.28}$$

Exercise

3.7. Explain these equations.

3.8.7 Model Improvement

Let string S of length n be observed as the emitted string. Based on S we compute the variables as above and use them to update transition and emission probabilities as follows, for states q and r different from initial state q_0:

$$
T^{\text{update}}(q, r) = \frac{\sum_{j=1}^{n-1} \eta_j(q, r)}{\sum_{j=1}^{n-1} \gamma_j(q)}
$$

$$
T^{\text{update}}(q_0, q) = \gamma_1(q) \tag{3.29}
$$

$$
E^{\text{update}}(q, x) = \frac{\sum_{i \leq n, S(i)=x} \gamma_i(q)}{\sum_{i \leq n} \gamma_i(q)}
$$

It can be shown (see [4]) that this update is a particular case of expectation maximization, and as such indeed leads to improvement of model likelihood. Iterated application thus approaches a local maximum of the model likelihood function.

3.9 Expressiveness of Viterbi Equations

The Viterbi algorithm was a simple application of the general principle of dynamic programming. It is interesting to see that often applications of dynamic programming may be reinterpreted as Viterbi equations for a suitably designed Hidden Markov model. This shows, in a certain sense, a certain generality of the Hidden Markov approach. We illustrate this by showing how affine gap alignment treated in Sect. 3.3 may be reinterpreted as Viterbi algorithms of a suitably designed Hidden Markov model (we follow the treatment in [27]).

The application of the gap alignment equations described in Sect. 3.3 may be visualized by the graph shown in Fig. 3.4 containing start node q_0, and three further nodes called 'a', 'b', and 'c' that correspond to the three functions optimized. Index pair (i, j) maintains which one of the characters of the strings S and T, namely $S(i)$ and $T(j)$, are the actual characters that are next to be aligned or aligned with a spacing symbol. Each node 'a', 'b', and 'c' contains information on how these indices are to be updated whenever the corresponding node is visited. Links contain scores that are to be added onto a growing score whenever the corresponding link is traversed. Note that there are no links between nodes 'b' and 'c'; this expresses our convention that any two b- and c-blocks must be separated by at least one a-block.

This graph gives rise to a similar structure of a Hidden Markov model (HMM) designed for the emission of an alignment (Fig. 3.5). First, the proposed HMM has states q_0, A, B, and C. State A emits pairs (x, y), state B emits pairs $(-, y)$, and state C emits pairs $(x, -)$, with characters x and y from alphabet Σ. Correspondingly, emission probabilities are called $P(x, y)$, $P(x)$, and

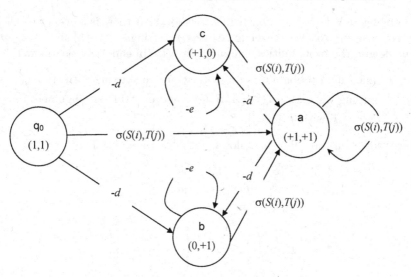

Fig. 3.4. Gap alignment modelled by a graph

$P(y)$. Assume they are taken from known sequence alignments by extracting frequencies. There are state transition probabilities ϵ and δ that occur at corresponding positions where numbers $-e$ and $-e$ occurred in Fig. 3.4. Note that some of the transition probabilities are thus fixed because all probabilities at outgoing links of a node must sum up to value 1. So far, there is only a somehow superfluous similarity between both diagrams.

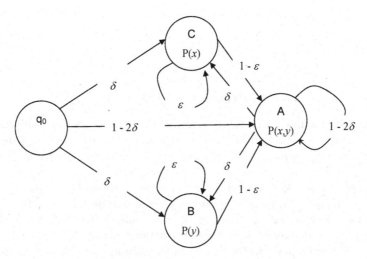

Fig. 3.5. Hidden Markov model for gap alignment

Looking at Viterbi equations will soon show that there is a deeper relation between both approaches. Given fixed observed strings $S[1 \ldots n]$ and $T[1 \ldots m]$ let us denote the probabilities computed by Viterbi equations as follows:

Q_{ij} =maximum probability of a state sequence starting with state q_0, ending with state $Q \in \{A, B, C\}$, and emitting an alignment of $S[1 \ldots i]$ with $T[1 \ldots j]$.

Viterbi equations now look as follows, with $i > 1$ and $j > 1$:

$$A_{1,1} = P(S(1), T(1))(1 - 2\delta)$$
$$A_{i,1} = P(S(i), T(1))(1 - \epsilon)C_{i-1,0}$$
$$A_{1,j} = P(S(1), T(j))(1 - \epsilon)B_{0,j-1}$$
$$A_{i,j} = P(S(i), T(j)) \max \begin{cases} (1 - 2\delta)A_{i-1,j-1} \\ (1 - \epsilon)B_{i-1,j-1} \\ (1 - \epsilon)C_{i-1,j-1} \end{cases}$$
$$B_{0,1} = P(T(1))\delta$$
$$B_{0,j} = P(T(j))\epsilon B_{0,j-1}$$
$$B_{1,1} = 0$$
$$B_{i,1} = 0 \tag{3.30}$$
$$B_{1,j} = P(T(j)) \max \{\delta A_{1,j-1}, \epsilon B_{1,j-1}\}$$
$$B_{i,j} = P(T(j)) \max \{\delta A_{i,j-1}, \epsilon B_{i,j-1}\}$$

$$C_{1,0} = P(S(1))\delta$$
$$C_{i,0} = P(S(i))\epsilon C_{i-1,0}$$
$$C_{1,1} = 0$$
$$C_{1,j} = 0$$
$$C_{i,1} = P(S(i)) \max \{\delta A_{i-1,1}, \epsilon C_{i-1,1}\}$$
$$C_{i,j} = P(S(i)) \max \{\delta A_{i-1,j}, \epsilon C_{i-1,j}\}$$

Note that probabilities with value 0 express the convention that b-blocks and c-blocks must be separated by at least one a-block. Next we take care that the terms within the max in the equation for $A_{i,j}$ get the same factor $(1 - 2\delta)$. This is done by replacing factor $(1 - \epsilon)$ from terms $B_{i-1,j-1}$ and $C_{i-1,j-1}$ with factor $(1 - 2\delta)$ and observing that in the execution of equations defining $B_{i-1,j-1}$ and $C_{i-1,j-1}$, we must sooner or later switch back to an A-term again (otherwise we would end with one of the impossible cases having probability 0). When this happens we introduce the missing factor $(1 - \epsilon)/(1 - 2\delta)$. By the same way we replace factor $(1 - \epsilon)$ in the second and third equation with $(1 - 2\delta)$ and compensate this by introducing a further factor $(1 - \epsilon)/(1 - 2\delta)$ in the definition of $C_{1,0}$ and $B_{0,1}$ (note that computations of $A_{i,1}$ and $B_{1,j}$ - and only these - finally end with these terms).

$$A_{1,1} = P(S(1), T(1))(1 - 2\delta)$$
$$A_{i,1} = P(S(i), T(1))(1 - 2\delta)C_{i-1,0}$$
$$A_{1,j} = P(S(1), T(j))(1 - 2\delta)B_{0,j-1}$$
$$A_{i,j} = P(S(i), T(j))(1 - 2\delta) \max \{A_{i-1,j-1}, B_{i-1,j-1}, C_{i-1,j-1}\}$$
$$B_{0,1} = P(T(1))\delta \frac{(1 - \epsilon)}{(1 - 2\delta)}$$
$$B_{0,j} = P(T(j))\epsilon B_{0,j-1}$$
$$B_{1,1} = 0$$
$$B_{i,1} = 0$$
$$B_{1,j} = P(T(j)) \max \left\{ \delta \frac{(1 - \epsilon)}{(1 - 2\delta)} A_{1,j-1}, \epsilon B_{1,j-1} \right\}$$
$$B_{i,j} = P(T(j)) \max \left\{ \delta \frac{(1 - \epsilon)}{(1 - 2\delta)} A_{i,j-1}, \epsilon B_{i,j-1} \right\}$$
$$C_{1,0} = P(S(1))\delta \frac{(1 - \epsilon)}{(1 - 2\delta)}$$
$$C_{i,0} = P(S(i))\epsilon C_{i-1,0}$$
$$C_{1,1} = 0$$
$$C_{1,j} = 0$$
$$C_{i,1} = P(S(i)) \max \left\{ \delta \frac{(1 - \epsilon)}{(1 - 2\delta)} A_{i-1,1}, \epsilon C_{i-1,1} \right\}$$
$$C_{i,j} = P(S(i)) \max \left\{ \delta \frac{(1 - \epsilon)}{(1 - 2\delta)} A_{i-1,j}, \epsilon C_{i-1,j} \right\}.$$

$$(3.31)$$

The next adaptation to the equations of gap alignment concerns division by the following constant value that (as a constant) does not influence optimization on state sequences:

$$\prod_{i=1}^{n} P(S(i)) \prod_{j=1}^{m} P(T(j)). \qquad (3.32)$$

Note that this value measures probability of independent occurrence of strings S and T. Division by these values is not at the end of the complete recursion, but "on the fly" at suitably chosen moments of the computation. This leads to the following formulas:

$$A_{1,1} = \frac{P(S(1), T(1))}{P(S(1))P(T(1))}(1 - 2\delta)$$

$$A_{i,1} = \frac{P(S(i), T(1))}{P(S(i))P(T(1))}(1 - 2\delta)C_{i-1,0}$$

$$A_{1,j} = \frac{P(S(1), T(j))}{P(S(1))P(T(j))}(1 - 2\delta)B_{0,j-1}$$

$$A_{i,j} = \frac{P(S(i), T(j))}{P(S(i))P(T(j))}(1 - 2\delta) \max \begin{cases} A_{i-1,j-1} \\ B_{i-1,j-1} \\ C_{i-1,j-1} \end{cases}$$

$$B_{0,1} = \delta \frac{(1 - \epsilon)}{(1 - 2\delta)}$$

$$B_{0,j} = \epsilon B_{0,j-1}$$

$$B_{1,1} = 0$$

$$B_{i,1} = 0$$

$$B_{1,j} = \max \left\{ \delta \frac{(1 - \epsilon)}{(1 - 2\delta)} A_{1,j-1}, \epsilon B_{1,j-1} \right\}$$

$$B_{i,j} = \max \left\{ \delta \frac{(1 - \epsilon)}{(1 - 2\delta)} A_{i,j-1}, \epsilon B_{i,j-1} \right\}$$

$$C_{1,0} = \delta \frac{(1 - \epsilon)}{(1 - 2\delta)}$$

$$C_{i,0} = \epsilon C_{i-1,0}$$

$$C_{1,1} = 0$$

$$C_{1,j} = 0$$

$$C_{i,1} = \max \left\{ \delta \frac{(1 - \epsilon)}{(1 - 2\delta)} A_{i-1,1}, \epsilon C_{i-1,1} \right\}$$

$$C_{i,j} = \max \left\{ \delta \frac{(1 - \epsilon)}{(1 - 2\delta)} A_{i-1,j}, \epsilon C_{i-1,j} \right\}.$$

(3.33)

Finally, we replace the value with their logarithms. This does not influence optimization as the logarithm is a monotone function. Products simply turn into sums. Also note that the logarithm of a max is, of course, equal to the max of logarithms. Defining $a_{i,j} = \log(A_{i,j})$, $b_{i,j} = \log(B_{i,j})$, $c_{i,j} = \log(C_{i,j})$, the equations below are obtained. Within these equations we used the following abbreviations:

$$\sigma(x, y) = \log \frac{P(x, y)}{P(x)P(y)} + \log(1 - 2\delta)$$

$$d = -\log \frac{\delta(1 - \epsilon)}{(1 - 2\delta)}$$

$$e = -\log \epsilon.$$

(3.34)

Note that $\sigma(x, y)$ exactly corresponds (up to a constant additive shift that does not influence quality of a scoring function) to the sort of scoring functions used in practice. Gap parameters are plausible, too. Consider value e. The smaller the probability ϵ for running into gap generating state B or C, the stronger is a prolongation of a gap penalized. Interpretation of gap opening term $-d$ is not as clear since it is a more complicate mixture of probabilities.

$$
\begin{aligned}
a_{1,1} &= \sigma(S(1), T(1)) \\
a_{i,1} &= \sigma(S(i), T(1)) + a_{i-1,0} \\
a_{1,j} &= \sigma(S(1), T(j)) + b_{0,j-1} \\
a_{i,j} &= \sigma(S(i), T(j)) + \max\left\{a_{i-1,j-1}, b_{i-1,j-1}, c_{i-1,j-1}\right\} \\
b_{0,1} &= -d \\
b_{0,j} &= -e + b_{0,j-1} \\
b_{1,1} &= -\infty \\
b_{i,1} &= -\infty \\
b_{1,j} &= \max\left\{-d + a_{1,j-1}, -e + b_{1,j-1}\right\} \\
b_{i,j} &= \max\left\{-d + a_{i,j-1}, -e + b_{i,j-1}\right\} \\
c_{1,0} &= -d \\
c_{i,0} &= -e + c_{i-1,0} \\
c_{1,1} &= -\infty \\
c_{1,j} &= -\infty \\
c_{i,1} &= \max\left\{-d + a_{i-1,1}, -e + c_{i-1,1}\right\} \\
c_{i,j} &= \max\left\{-d + a_{i-1,j}, -e + c_{i-1,j}\right\}
\end{aligned}
\tag{3.35}
$$

These are exactly the equations describing affine gap alignment with scoring function $\sigma(x, y)$ and gap function $f(k) = -d - (k - 1)e$. Thus we have shown that letting run the HMM designed above can indeed be interpreted as executing the dynamic programming equations of affine gap alignment with a natural definition of scoring and gap penalizing. Conversely, every affine gap alignment approach with a scoring function that can be written as log-odds ratio of a suitable distribution on character pairs is realized by a HMM of the sort described above.

3.10 Lifted Phylogenetic Alignment

3.10.1 Problem Restated

Given a list of present species $S_1, S_2, \ldots, S_n$ together with a tree structure T expressing evolutionary history of these species from ancestors (thus $S_1, S_2, \ldots, S_n$ appear at the leaves of the tree), we search for an assignment of strings to inner nodes such that the sum of alignment scores of connected pairs of strings is maximized. Such an assignment would thus be an optimal guess of how the ancestors of $S_1, S_2, \ldots, S_n$ looked like. As Chap. 5 will show, this problem is NP-hard even for the simple looking special case of a tree consisting only of a root and n leaves (also called a "star"). The chapter on approximation algorithms will demonstrate that there is always an assignment of strings to inner nodes of the tree which uses only the given leaf strings $S_1, S_2, \ldots, S_n$ as ancestors. Such an assignment will be called a "lifted assignment". Stated in biological terms, a lifted assignment makes the (surely unrealistic) assumption that there was no extinction of species during evolution. A dynamic programming approach readily computes an optimal lifted assignment (see [31]). Let us denote by $\pi_T(S_1, \ldots, S_n)$ the maximum possible sum of alignment scores between connected strings, taken over all lifted phylogenetic alignments of tree T.

LIFTED PHYLOGENETIC ALIGNMENT
Given rooted tree with an assignment of strings $S_1, S_2, \ldots, S_n$ to its leaves, compute $\pi_T(S_1, \ldots, S_n)$.

3.10.2 Parameterization and Conditioning

For an arbitrary node v of tree T (v may also be a leaf) consider the subtree $T(v)$ below (and including as root) node v. For a fixed string L occurring as the assignment to a leaf of subtree $T(v)$, we consider all lifted phylogenetic assignments for subtree $T(v)$ with the additional constraint that string L is assigned to root v. Thus the conditioning associated with string L expresses which of the leaf strings of $T(v)$ is the common ancestor of all other leaf strings. Denote by $\pi(v, L)$ the optimal value of a phylogenetic alignment of subtree $T(v)$ with the restriction that root v is assigned string L. Having computed all values $\pi(v, L)$ the original problem is solved as follows with r being the root of tree T:

$$\pi_T(S_1, \ldots, S_n) = \max_{1 \leq k \leq n} \pi(r, S_k). \tag{3.36}$$

3.10.3 Bellman Principle

In an optimal lifted phylogenetic alignment of a tree T with assignment of string L to the root of T, all of its subtrees must obviously also be optimal lifted phylogenetic alignments, though eventually with assignments to their root nodes with strings different from string L.

3.10.4 Recursive Solution

For a leaf node v with leaf inscription L we obviously have $\pi(v, L) = 0$. Now consider a non-leaf node v with successor nodes $v_1, v_2, \ldots, v_m$. Assume that numbering is such that L occurred as leaf description within subtree $T(v_1)$. subtree $T(v_1)$ thus must contribute to an optimal phylogenetic alignment value $\pi(v_1, L)$ and value 0 for the link between nodes v and v_1 (note that we assigned the same string L to both), whereas subtree $T(v_i)$, for $i > 1$, contributes the maximum possible value $\pi(v_i, L_i) + \sigma_{\mathrm{opt}}(S, L_i)$, with L_i ranging over all inscriptions at leaves from subtree $T(v_i)$.

$$\pi(v, L) = \pi(v_1, L) + \sum_{i=2}^{m} \max_{L_i} \pi(v_i, L_i) + \sigma_{\mathrm{opt}}(L, L_i) \qquad (3.37)$$

3.10.5 Counting and Overall Complexity

Tree T possesses at most n inner nodes (usually much less than the number n of leaves). Thus there are $O(n^2)$ terms $\pi(v, L)$ to be computed. Assume that strings $S_1, S_2, \ldots, S_n$ are bounded in length by k. Every computation of a new value from already computed ones according to the recursive equation above requires $O(n^2 k^2)$ steps due to summation over at most n successor nodes, within each sum a max computation over at most n leaf strings, and within each max computation the calculation of an optimal alignment between strings of length k. Thus, overall complexity is $O(n^4 k^2)$.

3.10.6 Preprocessing of Simple Terms

Preprocessing optimal alignments between any two leaf strings in time $O(n^2 k^2)$ saves execution time since we can simply call within each max term those preprocessed optimal alignment values leading to an improvement of overall complexity to $O(n^4 + n^2 k^2)$.

3.11 Protein Threading

3.11.1 Problem Restated

The ingredients of protein threading as introduced in Sect. 2.10.2 are: a protein S of length n with unknown tertiary structure, a known protein T with *loop-core segmentation* $T = L_0, C_1, L_1, C_2, \ldots, C_{m-1}, L_{m-1}, L_m$, strongly conserved *core segments lengths* $c_1, c_2, \ldots, c_m$, *loop segments lengths* that are assumed to be variable within certain limits given by $\lambda_i \leq l_i \leq \Lambda_i$, and a scoring function for threadings $t_1, \ldots, t_n$ of the simple form:

$$f(t_1, \ldots, t_n) = \sum_{p=1}^{m} g(t_p). \qquad (3.38)$$

Here, term $g(t_p)$ is an abbreviation for some function $g(C_p, C^p(t_p))$ of the core segment C_p in T and substring $C^p(t_p)$ of S of length c_p starting at position t_p. An example might be

$$g(t_p) = d_{\mathrm{opt}}(C_p, C^p(t_p)) \text{ with } C^p(t_p) = S[t_p \ldots t_p + c_p - 1] \qquad (3.39)$$

with a distance function d. The main point is that we discard any interactions between different core segments.

3.11.2 Parameterization and Conditioning

Instead of looking at the whole string S of length n, we consider prefixes $S[1\ldots j]$. Moreover, instead of considering the whole string T we take prefix $T\langle i \rangle = L_0, C_1, L_1, C_2, \ldots, C_i, L_i$ with loop-core segmentation up to C_i, L_i. We compute the following terms $F(i,j)$ that are defined as the minimum value

$$f(t_1, \ldots, t_i) = \sum_{p=1}^{m} g(t_p). \qquad (3.40)$$

taken over all threadings $t_1, \ldots, t_i$ of $S[1\ldots j]$ into the loop-core structure of $T\langle i \rangle$ up to loop segment L_i.

3.11.3 Bellman Principle

... is obviously fulfilled.

3.11.4 Recursive Solution

We start with the case $i = 0$ and $j \in [\lambda_0 \ldots \Lambda_0]$. This means that only loop segment L_0 is available for mapping $S[1\ldots j]$. Since $j \in [\lambda_0 \ldots \Lambda_0]$ was assumed mapping can be done. Costs are zero, as nothing is summed up. Thus $F(0,j) = 0$.

Next we treat the case $i = 0$ and $j \notin [\lambda_0 \ldots \Lambda_0]$. Here no threading exists that fulfils the length constraints. We express this by defining $F(0,j) = \infty$.

Now we treat the case $i > 0$ and $1 \leq j - c_i - \lambda_i + 1$. Whatever we choose as start position t_i for core segment of length c_i in $S[1\ldots j]$, there must be at least $c_i + \lambda_i$ characters between positions t_i and j available for a realization of ith core segment of length c_i and i^{th} loop segment of length at least λ_i. This constrains t_i to $t_i \leq \lambda(i,j) = j - c_i - \lambda_i + 1$. As $1 \leq j - c_i - \lambda_i + 1$ was assumed there is at least one choice for t_i possible. As conversely loop segment numbered i must have length at most Λ_i this constrains t_i to $t_i \geq \Lambda(i,j) = j - c_i - \lambda_i + 1$. This leads to the following formula.

$$F(i,j) = \min_{\Lambda(i,j) \leq t_i \leq \lambda(i,j)} g(t_i) + F(i-1, t_i - 1) \qquad (3.41)$$

Finally we treat the case $i > 0$ and $1 > j - c_i - \lambda_i + 1$. Here, no admissible threadings are possible, thus $F(i,j) = +\infty$.

3.11.5 Counting and Overall Complexity

There are mn terms $F(i,j)$ to be computed. Each min computation takes $O(m)$ steps. This leads to an overall complexity of $O(mn^2)$.

3.12 Bibliographic Remarks

The principle of dynamic programming is attributed to Bellman [8, 9]. Any of the standard textbooks Clote & Backofen [20], Gusfield [31], Pevzner [64], Setubal & Meidanis [68], and Waterman [78] are full of applications of dynamic programming.

4

Intelligent Data Structures

4.1 Representation Matters

We start the chapter on intelligent data structures and the enormous impact suitably chosen representations have on transparency and efficiency of solutions with a couple of examples that indicate what is intended within this chapter.

4.1.1 Intelligent Data Structures Used in Computer Science

Computer Science is full of properly designed data structures that support important applications, sometimes in a rather astonishing extend going far beyond what is achievable with more naive approaches. Prominent key words are sorted arrays, various sorts of balanced trees, Fast Fourier Transform, to name only a few. The value of properly designed data structures and representations is already observed on much simpler levels, for example with respect to integer arithmetic. As it is well known, Arabic numerals support equally well operations of addition and multiplication, whereas integer notation used by ancient Romans is not comparably useful. Not only a proper representation of operating data may considerably influence complexity, proper representations also play an important role in making complex constructions and mathematical proofs readable. Numerous examples can be found in every mathematics textbook. Finally, suitable visualizations often make things from daily life understandable, whereas improper representations may hinder understanding or even may lead to erroneous imaginations. This latter phenomenon will be illustrated with a few examples in the following sections.

4.1.2 Covering a Truncated Board

Consider an 8×8 board with the upper-left and lower-right field cut. Can you cover it using 31 dominos? Dominos may be placed horizontally and vertically, but are not allowed to overlap. Look at the diagrams in Fig. 4.1 and see which of the drawings immediately shows the answer to the posed questions, whereas the other one misleads the reader to try out lots of placing variants for the dominos.

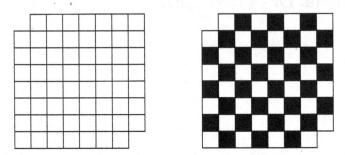

Fig. 4.1. Covering a truncated chess board

4.1.3 Choose Three Bits and Win

Two players A and B play the following game. They sit in front of a bit generator that will soon be started and will generate a stream of random bits with bits being independently drawn having equal probability for bit 0 as for bit 1. Player A has to choose a sequence of three bits. Player B can see what A has chosen and has to react with a different sequence of three bits (thus player B is not allowed to choose the same three bits that player A selected, though one or two of the chosen bits may coincide). Starting and watching the random bit generator, the player whose three bits are drawn in succession wins the game. For example, let A choose 001 and B choose 111. Observing bit stream 0101010000001 ... lets player A win. Which player has better chances to win? One might argue that A has eight choices for his sequence whereas B has only seven, what seems to favour player A. One might argue that every bit string is as good as every other, thus players A and B have identical winning chance. One might argue that player B reacts to what player A has chosen thus could select a somehow more favourable bit string than A did. But if this were the case one could argue that player A could then choose in advance this more favourable bit string himself, and in case that player B could further improve this choice, could again choose the more favourable string. Finally, A had chosen the most favourable string at all.

The visualization of the working of the random bit generator shown in Fig. 4.2 makes the game transparent. Here, we use a graphical representation with four nodes labelled 00, 01, 10, 11, and eight directed links each labelled either 0 or 1. Each node represents what were the last two generated bits, whereas outgoing link represents the next generated bit. Note that links always point

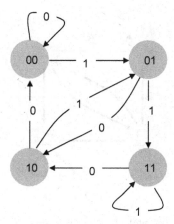

Fig. 4.2. Random bit generator

to the intended target node. Now the working of the random bit generator can be seen as choosing a particular edge (that edge corresponds to the first three generated bits), and then walking through the graph. By the same way, the selection of three bits each player did can be seen as selection of a particular edge by each player. The following graphs immediately show how B must react to any choice of A in order to have a better chance to win. Furthermore, it is easy to estimate how much higher chances are for B to win than for A. We indicate three qualitatively different choices A can make and a corresponding better choice for B. The possibility for player B to outperform every selection player A makes is no contradiction to the fact that there are only eight possible choices for A as the relation of "being better" is not a linear ordering though its naming suggests exactly this.

In the diagrams of Fig. 4.3, labels of nodes and edges do not play any role, thus they are omitted. Bold links indicate which drawings of the first three bits with certainty lead to a win of player B, and by broken links we indicate which drawings of the first three bits lead to at least 50% winning chance for player B. The remaining links either favour player A or do not uniquely favour one of the players (thus these are ambivalent links). For simplicity we assume that there are no infinite bit streams generated with neither a win for A nor B.

4.1.4 Check a Proof

All of the algorithmic problems treated within this book are search problems that draw their complexity from exponential size of their solution search spaces. For some of them a complete search for a solution through the search space with exponential size can be avoided using clever algorithms, while others are shown to be NP-hard leaving no chance (unless P = NP) to find an efficient algorithm. Whereas searching for a solution is the algorithmic core

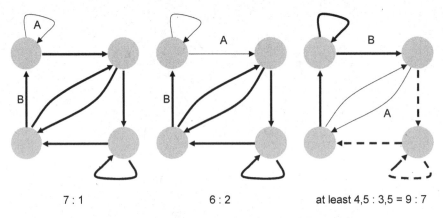

7 : 1   6 : 2   at least 4,5 : 3,5 = 9 : 7

Fig. 4.3. Estimating winning chance of B over A

of these problems, checking a proposed solution for correctness is the simple part requiring "only" polynomial time (in most cases quadratic time or even better). Nevertheless, checking correctness of a solution normally requires at least reading of the complete solution, thus time linear in the length of the solution. Only for special cases considerable parts of a solution have no influence on their correctness and thus can be left out. As an example, checking if a bit string of length n is a satisfying truth-value assignment for a Boolean formula with n variables, in general requires reading of the whole bit string. If we relax certainty of the outcome of a correctness test to high confidence of the outcome, things radically change. Let us first fix what is expected with respect to confidence of correctness checks. In case that a satisfying truth-value assignment is checked we expect with certainty a positive answer "satisfies the formula". In case that a non-satisfying assignment is checked we expect with some fixed confidence $\delta < 1$ (for example 99%) a negative answer "does not satisfy the formula". Tolerating such almost correct answers, checking solutions may be dramatically accelerated. Assuming that solutions are not given as bit strings, but represented in a certain algebraic format based on linear function, primes, and polynomials, we must only randomly draw a (small) constant number c of bits of such a represented solution, then do a simple check of how these drawn bits are related to the Boolean formula which is to be tested for being satisfiable, and return the answer which with confidence δ is the correct one. The astonishing point is that the number of bits to be randomly drawn is fixed, i.e. a constant, independent of how long the Boolean formula or the proposed solution is. Metaphorically expressed, fixing a certain format for doctoral theses you might evaluate correctness of arbitrary such theses with, say 99% confidence, by always reading only, say 104 random drawn bits and checking these, regardless of how long the theses are. This is the content of one of the most beautiful and astonishing recent theorems of theoretical computer science with enormous applications concerning optimality of approximation algorithms, called the PCP theorem ("probabilistically checkable proofs") that was proved in the 1990s (see [2]). One is tempted to speculate about the question whether nature might also have invented such advantageous codes in the regulation of molecular processes.

4.2 PQ-Trees

4.2.1 Basic Notions

We described genome mapping based on hybridization experiments using sequence tagged sites as a fundamental task in genome analysis, leading to the CONSECUTIVE ONES problem (Sect. 2.2.1). Given a binary matrix M with n rows and m columns, we attempt to determine all permutations of the columns of M that lead to a matrix with consecutive bits 1 in every row. The process of successively working towards this goal by suitable rearrangements of columns is best presented using PQ-trees which were introduced in Sect. 2.2.1. We now describe the details of this rearrangement process. We start with a tree T_0 consisting of a P-node as root and m leaves (Fig. 4.4). Note that T_0 is the most liberal tree representing arbitrary orderings of the columns.

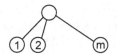

Fig. 4.4. Initial PQ-tree

Now we successively reorganize this tree as moderate as possible such that for every row consecutiveness of its bits 1 is enforced. Assume this has been done for rows $1, \ldots, i$ leading to tree T_i. Next we consider row $i + 1$.

- Unmark all nodes of tree T_i (markings left over from iteration before).
- Mark every leaf that corresponds to a column with bit 1 in row $i + 1$.
- Determine the deepest node r with all marked leaves below node r.

Note that the goal of the actual iteration is to make all marked leaves consecutive. This is done in a recursive manner using certain rules that appropriately reorganize the tree below the actual working node assuming that all of its subtrees have already been successfully reorganized. Note that "subtree of some node" is always understood as subtree immediately below that node. In the following rules it will be important to distinguish whether we are actually working at node r or at a node q different from node r. Working at node r means that there will be no further marked leaves other than those below r that must be considered, whereas working at a different node q means that further marked leaves come up later. This influences the process of restructuring T_{i+1} at the actual working node. In the description of this process we use the following notational conventions.

- A subtree is called "full" if all of its leaves are marked.
- A subtree is called "empty" if none of its leaves are marked.
- A subtree is called "partial" if it is neither full nor empty.

We always indicate full trees by black, empty trees by white, and partial trees by a mixed black-and-white shading. In the diagrams below, two full resp. empty subtrees drawn side by side always stand for a finite number (possibly zero) of subtrees. Whenever a minimum number of subtrees of a certain sort is required for a case, this is explicitly indicated by a lower bound written below the subtrees of that sort.

Fig. 4.5. Tree: **(a)** full; **(b)** empty; **(c)** partial. Finite number of **(d)** full or **(e)** empty trees; at least k **(f)** full or **(g)** empty trees

After having processed any node $q \neq r$ by one of the rules described below the following *invariant properties* will be fulfilled (see Fig. 4.6); note that for root node r this does not necessarily hold:

- If q is a P-node then the subtree with root q is either full or empty.
- If q is a Q-node then q has only full and empty subtrees, but no partial subtrees. Furthermore, all of its full subtrees, as well as all of its empty subtrees are arranged consecutively.

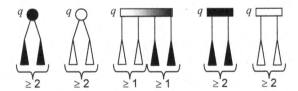

Fig. 4.6. Only trees like the ones shown occur in the execution of the algorithm

In the description of the rules below we assume that before rule application invariance properties hold, and easily verify that after rule application invariance properties again hold. We may always assume that the actually processed node has at most two partial subtrees as otherwise the marked leaves could not be made consecutive (see Fig. 4.7).

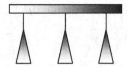

Fig. 4.7. More than two partial subtrees below Q-node cannot be made consecutive

4.2.2 Transformation rules for marked PQ-Trees

The rules to be described next distinguish whether actual working node

- is a P-node or a Q-node,
- coincides with node r or is a node q different from r,
- has zero or one or two partial subtrees,
- has zero or one or more than one full resp. empty subtrees.

The requirement for the second distinction stems from the fact that outside the subtree with root r no further marked leaves are to be considered; that means that subtrees may be arranged more liberally, whereas strictly below r subtrees may be arranged only at the left or right border of Q-nodes. Note that in the drawings below any tree stands for a whole collection of equivalent trees (arbitrary permutation of subtrees below a P-node, inversion of order of subtrees below a Q-node). We begin with the cases that the actual working node is a P-node.

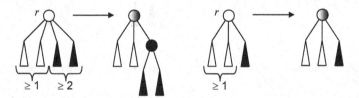

Fig. 4.8. Working at P-node r: no partial subtrees, at least one empty subtree, at least two resp. exactly one full subtree

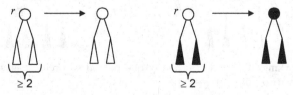

Fig. 4.9. Working at P-node r: no partial subtrees, only empty subtrees resp. only full subtrees

Fig. 4.10. Working at P-node q: no partial subtrees, at least two resp. exactly one empty subtree, at least two full subtrees

Fig. 4.11. Working at P-node q: no partial subtrees, at least two resp. exactly one empty subtree, exactly one full subtree

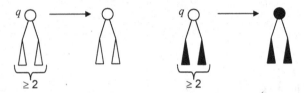

Fig. 4.12. Working at P-node q: no partial subtrees, only empty subtrees resp. only full subtrees

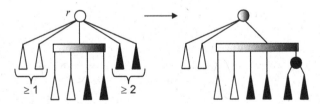

Fig. 4.13. Working at P-node r: exactly one partial subtree, at least one empty subtree, at least two full subtrees

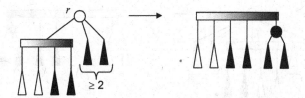

Fig. 4.14. Working at P-node r: exactly one partial subtree, no empty subtrees, at least two full subtrees

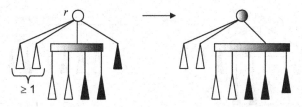

Fig. 4.15. Working at P-node r: exactly one partial subtree, at least one empty subtree, exactly one full subtree

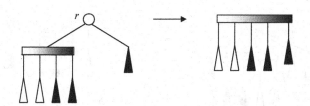

Fig. 4.16. Working at P-node r: exactly one partial subtree, no empty subtrees, exactly one full subtree

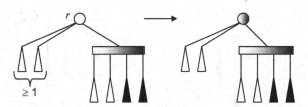

Fig. 4.17. Working at P-node r: exactly one partial subtree, at least one empty subtree, no full subtrees

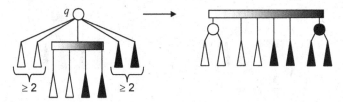

Fig. 4.18. Working at P-node q: exactly one partial subtree, at least two empty subtrees, at least two full subtrees

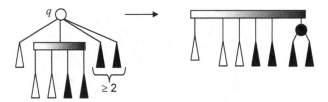

Fig. 4.19. Working at P-node q: exactly one partial subtree, exactly one empty subtree, at least two full subtrees

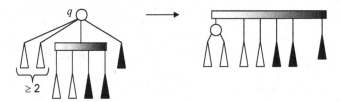

Fig. 4.20. Working at P-node q: exactly one partial subtree, at least two empty subtrees, exactly one full subtree

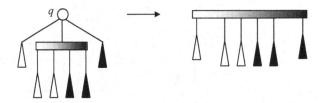

Fig. 4.21. Working at P-node q: exactly one partial subtree, exactly one empty subtree, exactly one full subtree

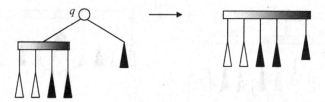

Fig. 4.22. Working at P-node q: exactly one partial subtree, no empty subtrees, exactly one full subtree

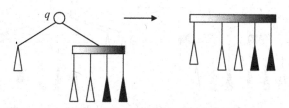

Fig. 4.23. Working at P-node q: exactly one partial subtree, exactly one empty subtree, no full subtrees

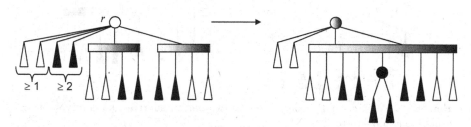

Fig. 4.24. Working at P-node r: exactly two partial subtrees, at least one empty subtree, at least two full subtrees

Note that in Fig. 4.24 as well as in the following four figures we are working at a node with two partial subtrees. We assume that we are working at node r since under the presence of two partial subtrees below some node $q \neq r$ marked leaves cannot be made consecutive.

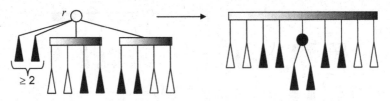

Fig. 4.25. Working at P-node r: exactly two partial subtrees, no empty subtrees, at least two full subtrees

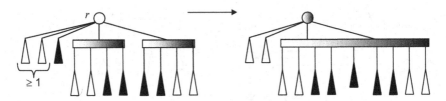

Fig. 4.26. Working at P-node r: exactly two partial subtrees, at least one empty subtree, exactly one full subtree

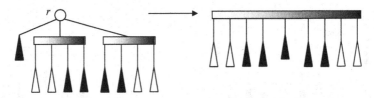

Fig. 4.27. Working at P-node r: exactly two partial subtrees, no empty subtree, exactly one full subtree

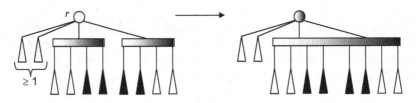

Fig. 4.28. Working at P-node r: exactly two partial subtrees, at least one empty subtree, no full subtrees

Next we treat the cases that occur when working node is a Q-node.

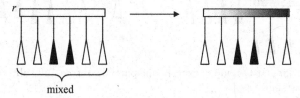

Fig. 4.29. Working at Q-node r: no partial subtrees, at least one empty subtree, at least one full subtree, full subtrees occurring consecutively

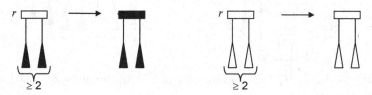

Fig. 4.30. Working at Q-node r: no partial subtrees, only full resp. only empty subtrees

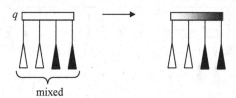

Fig. 4.31. Working at Q-node q: no partial subtrees, at least one empty subtree, at least one full subtree, full subtrees occurring consecutively either at the left or right border

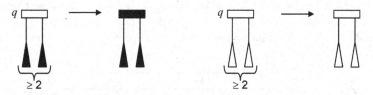

Fig. 4.32. Working at Q-node q: no partial subtrees, only full subtrees resp. only empty subtrees

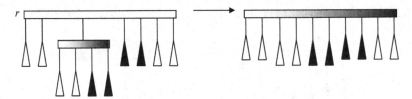

Fig. 4.33. Working at Q-node r: exactly one partial subtree, full subtrees occurring consecutively as shown

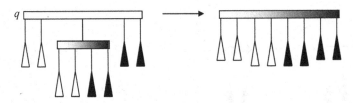

Fig. 4.34. Working at Q-node q: exactly one partial subtree, full subtrees occurring consecutively as shown

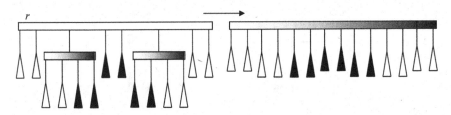

Fig. 4.35. Working at Q-node r: exactly two partial subtrees, full subtrees occurring consecutively as shown

Arriving at node r, the returned tree ideally looks as in Fig. 4.36 with a permutation $i_1, i_2, \ldots, i_m$. In that case, ordering of columns is uniquely determined up to complete inversion. In any other case we know that data either are inconsistent thus preventing an ordering of columns with consecutive ones, or are to less informative to fix a unique ordering of columns.

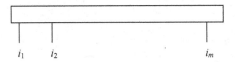

Fig. 4.36. "Almost" unique permutation

4.3 Suffix Trees

4.3.1 Outline of Ukkonen's Algorithm and Basic Procedures

Given string $T = T[1 \ldots n]$ of length n, Ukkonen's algorithm successively constructs suffix tree T_i for prefix string $T[1 \ldots i]$, with i increasing from 1 to n. Having constructed T_{i-1}, the construction of T_i is done by successively navigating to all positions in T_{i-1} that have as path label a suffix $T[j \ldots i-1]$ of $T[1 \ldots i-1]$, and then extending at this position suffix $T[j \ldots i-1]$ in a suitable manner by next character $T(i)$. This is done with j increasing from 1 to i. In particular, for $j = i$, navigation is to the root and thus the fresh suffix $T[i \ldots i]$ is realized as path label immediately below the root. Thus a rough description of the algorithm is as follows.

> initialize tree consisting only of a root node;
> **for** $i = 1$ to n **do**
>   **for** $j = 1$ to i **do**
>   set actual working position to the root;
>   navigate to position with path label $T[j \ldots i-1]$;
>   at that position insert next character $T(i)$;
>   **end for**
> **end for**

The process of navigating from the actual working position into a suffix tree along the characters of pattern P is abbreviated by $navigate(P)$. Details of an implementation are as follows. In case that P is the empty string we are done. Otherwise, we consider the first character c of P and check whether there is an edge starting at actual working node whose path label Q has first character c. If no such link is found, $navigate(P)$ returns "failure". If such an (unique) edge is found we further check whether Q is a prefix of P. If this is not the case, we again return "failure". If Q is found as prefix of P, say $P = QR$, we set actual working position to the target node of the edge with edge label Q and recursively proceed with $navigate(R)$. Obviously, calling $navigate(P)$ requires as many steps as P has characters. Knowing that P indeed is present in the considered suffix tree as the path label of a position allows a considerable acceleration of navigation. Having found the edge whose edge label Q has first character c (thus Q is a prefix of $P = QR$) we may readily jump without further checks to the target node of that edge and recursively proceed with $navigate(R)$. Thus, under knowledge that P indeed is present as path label, navigation along P requires only as many steps as nodes are visited along the path with path label P.

Besides *navigate* there will be two further procedures, called *skip* and *suffix*, which play an important role in the implementation of Ukkonen's algorithm. Having actual working position not at the root of the tree, *skip* goes upwards in the tree towards the next available node. Thus, sitting at a node that is not the root node, *skip* walks to its immediate predecessor node, whereas sitting between two characters of an edge label, *skip* walks to the source node of that edge label. Procedure *suffix* is applied only in case that actual working position is at a node that is neither the root nor a leaf, and then follows the unique suffix link (whose existence was proved in Chap. 2) starting at the actual working position, setting working position to the target node of this suffix link. Figure 4.37 summarizes and visualizes what happens when calling one of the procedures *navigate*(P), *skip*, or *suffix*.

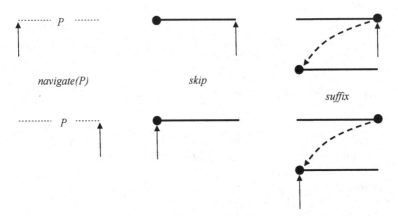

Fig. 4.37. Effect of procedures *navigate*(P), *skip*, *suffix*

4.3.2 Insertion Procedure

Now we describe in detail how the insertion of next character $T(i)$ at the position of insertion found by navigation takes place. Here we distinguish between four cases.

(1) Actual working position is at a leaf. Thus we know that actual suffix $T[j \ldots i-1]$ that is to be extended by next character $T(i)$ ends at the end of the path label of the edge leading to this leaf. Thus we simply have to append $T(i)$ to the edge label γ of the edge leading to the leaf. Setting new working position between γ and $T(i)$ will prove to be advantageous for later use (Fig. 4.38).

(2) Actual working position is at a node u that is not a leaf and next character $T(i)$ is not the first character of any edge label starting at node u. Then create a new leaf below u, mark it 'j' and label its edge with character $T(i)$ (Fig. 4.39). Actual working position is not changed.

(3) Actual working position is not at a node and character $T(i)$ is not the next symbol immediately right of the actual working position. Then create a new inner node u at the actual working position and a new leaf marked

Fig. 4.38. Insertion at an already constructed leaf

Fig. 4.39. Creation of a new leaf

'j' below this inner node, and label the edge between the two new nodes with $T(i)$. Furthermore, if u is not the root then initialize a dangling suffix link with source node u that will receive its target node in the next step when appending same character $T(i)$ to next prefix $T[j+1\ldots i-1]$. In case that $j > 1$ there was a dangling suffix link created one step before appending same character $T(i)$ to previous prefix $T[j-1\ldots i-1]$; take u as target node for this dangling suffix link. Actual working position is not changed, i.e. it now points to the fresh generated inner node (Fig. 4.40).

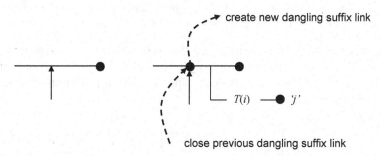

Fig. 4.40. Creation of a new inner node, new leaf, and new dangling suffix link

(4) Next character $T(i)$ is already present as the character immediately right of the actual working position. Then shift actual working position right to this occurrence of $T(i)$ (Fig. 4.41).

Fig. 4.41. Character already present

Note that only in case (4) the actual working position is changed. Letting actual working position in cases (1) - (3) unchanged will prove to be advantageous with respect to determining the next actual working position. The presence or creation of a suitable target node for each dangling suffix link at the step following the creation of a fresh dangling suffix link is shown next.

Lemma 4.1. *Dangling Suffix Links*

Whenever some new inner node u is created during insertion of next character $T(i)$ at the position with path label $T[j \ldots i-1]$, then $j < i$ holds, and insertion of same character $T(i)$ at the position with path label of the next suffix $T[j+1 \ldots i-1]$ either finds or creates a node v having path label $T[j+1 \ldots i]$. Node v thus is the correct target node for the dangling suffix link with source node u that was created in the step before.

Proof. A new inner node is created only in case (3). For $i = 1$ or $j = i$ there is no inner node created, thus $j < i$ holds. Case (3) applies whenever procedure $navigate(T[j \ldots i-1])$ found a position between two characters b and c of an edge label, with symbol c being different from $T(i)$. This means that $T[j \ldots i-1]c$ is a suffix of $T[1 \ldots i-1]$. Thus, $T[j+1 \ldots i-1]c$ is a suffix of $T[1 \ldots i-1]$, too. Thus, procedure $navigate(T[j \ldots i-1])$ finds a position with path label $T[j+1 \ldots i-1]$ such that symbol c also occurs immediately right of this position. If the position found is a node v, we are done. If the position found is not at a node, insertion procedure now creates according to case (3) a fresh node v with path label $T[j+1 \ldots i]$.   □

At this stage of discussion let us give a first estimation of how expensive the described algorithm is. Figure 4.42 indicates, for the case $n = 4$, all substrings of T that are successively traversed by the navigation procedure during development of the growing suffix tree.

	1	2	3	4
1	$T[1..1]$			
2	$T[1..2]$	$T[2..2]$		
3	$T[1..3]$	$T[2..3]$	$T[3..3]$	
4	$T[1..4]$	$T[2..4]$	$T[3..4]$	$T[4..4]$

Fig. 4.42. Outer and inner loop

For the case of arbitrary string length n, we navigate along one string of length n, along two strings of length $n-1$, along three strings of length $n-2$, and so on. This gives a rough estimation of running time by $1n^2 + 2(n-1)^2 + 3(n-2)^2 + \ldots + n = O(n^3)$. That is even worse than the naive $O(n^2)$ approach described in Chap. 2 for the construction of a suffix tree. Nevertheless, the algorithm introduced here offers several ways to improve running time.

4.3.3 Saving Single Character Insertions

Two observations show that only at most $2n$ of the $n(n+1)/2$ many entries of the table in Fig. 4.42 (for the case of arbitrary length n instead of length 4) must explicitly be visited during execution of the algorithm.

Lemma 4.2.
Whenever character $T(i)$ is inserted as edge label of a new created leaf immediately right of the position with label $T[j \ldots i-1]$, then all further characters $T(i+1), T(i+2), \ldots, T(n)$ will be appended via case (1) to the end of that same fresh leaf label.

Proof. This is obvious since, after having created leaf 'j' with path label $T[j \ldots i]$, further navigations to positions with path labels $T[j \ldots i], T[j \ldots i+ 1], \ldots, T[j \ldots n-1]$ always follow the same path towards just this new leaf 'j'. □

This allows us to append in a single step the whole string $T[i \ldots n]$ as label of the leaf marked 'j' at the time that leaf is created (with actual working position remaining immediately left of symbol $T(i)$) (Fig. 4.43).

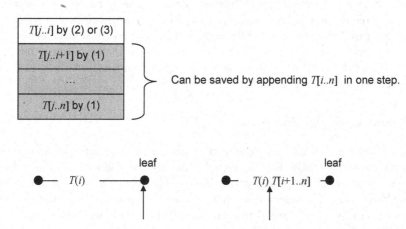

Fig. 4.43. Explicit insertion of all further characters can be saved

Lemma 4.3.
Whenever character $T(i)$ is found via case (4) to be already present immediately right of the working position with label $T[j \ldots i-1] = x\alpha$ (with a character x and string α), character $T(i)$ will be also found via case (4) to be present immediately right of working position with label $T[j+1 \ldots i-1] = \alpha$.

Proof. This is clear for the following reason. Under the assumptions above, successful navigation along $x\alpha T(i)$ means that $x\alpha T(i)$ occurs as prefix of a suffix of $T[1 \ldots i-1]$, hence $\alpha T(i)$ occurs as a prefix of a suffix of $T[1 \ldots i-1]$, too. Therefore, navigating along $\alpha T(i)$ also leads to a working position with label $\alpha T(i)$. $\qquad\square$

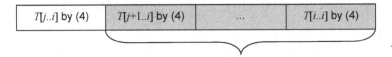

Steps can be saved since there is nothing to be done.

Fig. 4.44. No further suffixes must be visited

A typical run through the program above with the savings described looks as shown in Fig. 4.45 with saved entries indicated by shading (cases (2) and (3) in an entry lead to savings via case (1) in all entries below, case (4) in an entry leads to saving via case (4) right from the entry). What can be seen is that a linear trail through the table remains instead of the former quadratic traversal. Nevertheless, linear execution time is still not achieved. We are left with the problem to have at most $2n$ calls of procedure *navigate*, each with costs $O(n)$ to find the actual working position where next character has to be inserted. Instead of navigating at each step from scratch (i.e. starting at the root) to next actual working position, suffix links will help us to efficiently determine next actual working position from the previous one without necessarily having to go back to the root, thus speeding up navigation to constant time. Details are described in the next section.

4.3.4 Saving Navigation Steps

After having inserted or found the next character at a certain position in the diagram above, we show how to rapidly find the position of next insertion without doing unnecessary and costly navigation steps. There are a couple of simple situations. To understand what is done take a short look back to the construction above and see where actual working position was placed in cases (1) - (4).

(a) If we are working at the lower row $i = n$ with case (4) or at the lower-right corner $i = j = n$ with one of the cases (2), (3), or (4), the algorithm halts.
(b) If we are working at an inner row $i < n$ with case (4), actual working position just shifts one character to the right.
(c) If we are working at the rightmost field $j = i$ of an inner row $i < n$ with one of the cases (2) or (3), we know that insertion has been done immediately below the root, and again has to be done immediately below

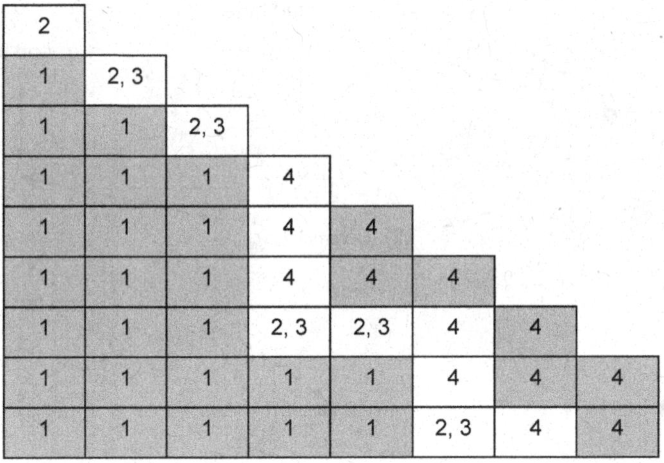

2							
1	2, 3						
1	1	2, 3					
1	1	1	4				
1	1	1	4	4			
1	1	1	4	4	4		
1	1	1	2, 3	2, 3	4	4	
1	1	1	1	1	4	4	4
1	1	1	1	1	2, 3	4	4

Fig. 4.45. Typical run with lots of savings

the root in the next step which works in the rightmost field $j + 1 = i + 1$ of the next row $i + 1$. Thus, actual working position is not altered.

(d) The only non-trivial situation is that we are working at an inner field with case (2) or (3). Assume that the path label of the former actual working position was $x\alpha$, with a single character x and some string α. What has to be found is the position with next suffix α as path label. This can be done without performing a complete navigation along α as follows (shown in Fig. 4.46). Calling procedure *skip*, we walk back to the next node strictly left of the actual working position. If we arrive at the root we must indeed navigate from scratch along suffix α to the next actual working position. Now assume that we arrived at some inner node u. Its path label is thus a prefix of $x\alpha$. Let us decompose $x\alpha$ into $x\beta\gamma$, with $x\beta$ being the path label of node u and γ being the label between node u and the former actual working position. Note that while skipping we have stored this label γ. Now following the suffix link that starts at node u leads us in one step to some node v with path label β. From node v we then navigate along remaining string γ. This finally leads us to the desired new actual working position with path label $\beta\gamma$.

4.3.5 Estimation of the Overall Number of Visited Nodes

Whereas execution of procedures *skip* and *suffix* requires constant time, the $O(n)$ many executions of procedures *navigate*(γ) seem to cause problems again. Though we always call procedure *navigate*(γ) only with a string γ that occurs right of the actual working position, and thus execution of *navigate*(γ) takes only as many steps as nodes are visited, and not as many as there are characters in γ, we must exclude the possibility that $O(n)$ nodes are visited

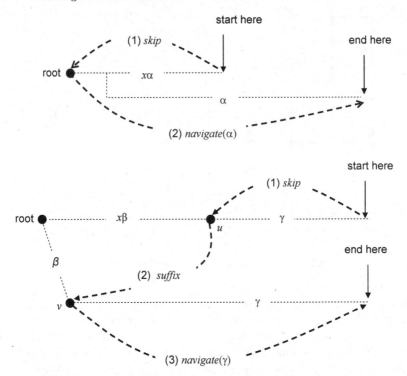

Fig. 4.46. Using suffix links for rapid navigation to next suffix

by the first *navigate* operation, that *skip* and *suffix* bring us back towards the root such that the second *navigate* operation again visits $O(n)$ nodes, and so on. This would, in summary, again lead to an overall estimation of $O(n^2)$ visited nodes for the complete algorithm execution, therefore nothing would be gained. Fortunately we can show that operation *suffix* will take us at most one node closer to the root, as it is trivially the case for operation *skip*.

Lemma 4.4.
Consider a suffix link leading from node u to node v. Then the path from the root down to u possesses at most one more node than the path from the root down to v. Thus, node v is at most one node closer to the root than node u is (measured in terms of nodes between the root and u and v, respectively).

Proof. Let the path label of node u be $x\alpha$ and the path label of node v be α, with a single character x. Consider a node w between the root and node u that is not the root. It has as path label a string $x\beta$ with a prefix β of α. Its suffix link thus points to a node $f(w)$ with path label β. Node $f(w)$ occurs on the path from the root to node v (and may be identical to the root). Obviously, function f is an injective function on the set of inner nodes occurring on the path from the root down to node u: different inner nodes w between the root and node u have different path labels, so do their images $f(w)$[1]. Hence, on the path between the root and node v there are at least as many nodes as there are inner nodes on the path between the root and node u.   □

In the examples shown in Fig. 4.47, there are five nodes on the path from the root down to node u, showing that the path from the root down to node v must have at least four nodes. It might well be that it possesses more than four nodes.

As a consequence, prior to each *navigate* step the preceding *skip* and *suffix* steps lead us at most two nodes closer to the root. This severely limits the possibility for too many node visits during *navigate* steps.

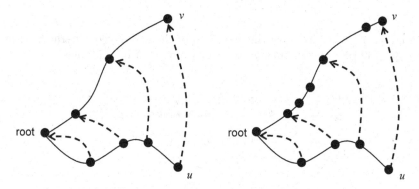

Fig. 4.47. Injective mappings on a path defined by suffix links

To estimate more precisely the overall number of visited nodes, define $g(i)$ to be the number of nodes that are visited by the algorithm in step i. After step i we have walked downwards over $g(1) + \ldots + g(i)$ nodes, and backwards over at most $2i$ nodes. There are least $g(1) + \ldots + g(i) - 2i$ nodes between the root and the actually visited node, thus after a further *skip* and *suffix* step at most $n - g(1) - \ldots - g(i) + 2i + 2$ many nodes available for the *navigate* step at stage $i + 1$. Hence, we obtain the following estimation:

$$g(i+1) \leq n - g(1) - \ldots - g(i) + 2(i+1) \, . \tag{4.1}$$

Stated differently, this leads to:

$$g(1) + \ldots + g(i) + g(i+1) \leq n + 2(i+1) \, . \tag{4.2}$$

Applying this with $i = n - 1$ gives the desired result:

$$g(1) + \ldots + g(n) \leq 3n \, . \tag{4.3}$$

---

[1] Note that the function f, as a function on the set of all inner nodes of a suffix tree, need not be injective. This can be seen in the example presented at the end of the chapter.

4.3.6 Ukkonen's Theorem

Theorem 4.5.
The suffix tree for a string of length n can be constructed in $O(n)$ steps.

We illustrate the working of the algorithm described above with the ananas$ example. We start with the single-root tree shown in Fig. 4.48.

Fig. 4.48. Step 0

Figures 4.49 to 4.57 show the evolution of the suffix tree. Characters that are explicitly placed without any savings are written in bold face.

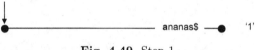

Fig. 4.49. Step 1

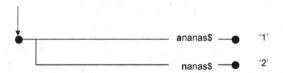

Fig. 4.50. Step 2

Fig. 4.51. Step 3

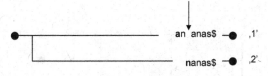

Fig. 4.52. Step 4

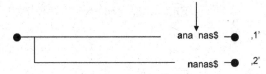

Fig. 4.53. Step 5

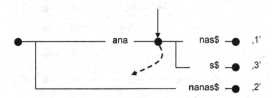

Fig. 4.54. Step 6

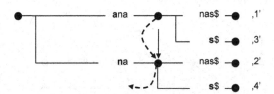

Fig. 4.55. Step 7

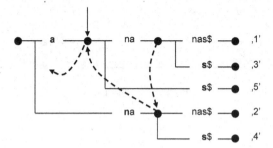

Fig. 4.56. Step 8

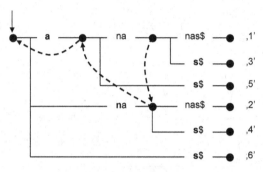

Fig. 4.57. Step 9

Figure 4.58 shows the final suffix tree. Figure 4.59 displays where savings took place and which cases occurred.

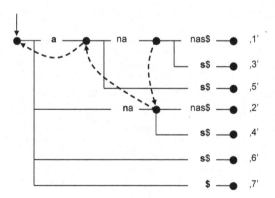

Fig. 4.58. Step 10 with final suffix tree

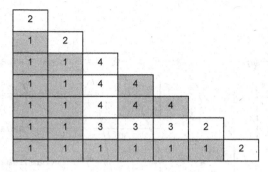

Fig. 4.59. Savings for the case of the ananas$ example

The example string T = ananas$ is too simple to lead to suffix link traversals during construction of the tree. Indeed, taking advantage of suffix links requires considerably longer and more involved strings as they typically occur in bioinformatics applications. As an example, containing suffix links that are indeed used during execution of Ukkonen's algorithm, the reader may build up the suffix tree for string T = abcdbcabcaacaaabaaaaabcda. The intermediate state after having arrived at a suffix tree for prefix abcdbcabcaacaaabaaaa is shown in Fig. 4.60. The reader may complete the construction by inserting the last 5 characters a, b, c, d, a, and see how often suffix links are traversed.

4.3.7 Common Suffix Tree for Several Strings

Having defined the notion of a suffix tree for single strings there are also applications requiring the *common suffix tree* for several strings. How to define such a common suffix tree is rather obvious. It should have all suffixes of all considered strings represented as path label of unique positions, ideally as path label of leaves that are also equipped with labels indicating which suffix of which strings is represented at that leaf. As the example below shows there may exist leaves whose path label is a suffix of several strings. The example considers two strings T = ananas$ and S = knasts$. Their common suffix tree is shown in Fig. 4.61 (markers 'T', 'S', and 'TS' at inner nodes are explained below). Ukkonen's algorithm is easily generalized to the case of more than one string. In a first run, construct the suffix tree for the first string T. In addition, mark all leaves with 'T' indicating that these represent suffixes of T. In a second run of Ukkonnen's algorithm also integrate all suffixes of S into a further growing tree and mark leaves that represent suffixes of S by 'S'. In addition, lift markings to inner nodes by marking inner node u with 'S' or 'T' or 'TS', in case that below u there are only leaves marked 'S' or only leaves marked 'T' or leaves marks 'S' as well as leaves marked 'T', respectively. It is obvious how to generalize definition and construction to the case of finitely many strings.

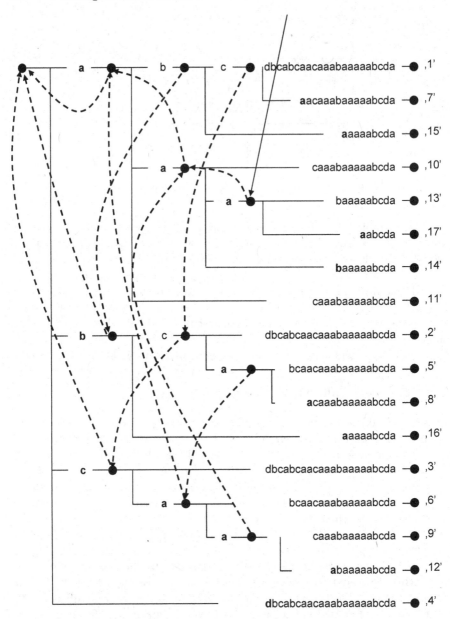

Fig. 4.60. The benefit of suffix links

4.3.8 Applications of Suffix Trees

The most complete collection of suffix tree applications is surely Gusfield [31].
We present here four examples of applications that drastically demonstrate
the usefulness of suffix trees. It is not wrong to state that without suffix trees

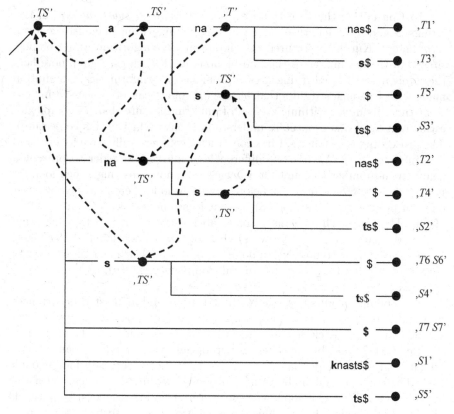

Fig. 4.61. Common suffix tree for strings ananas\$ and knasts\$

bioinformatics would not exists, at least not in the deeply established state it is today. The applications are:

(a) Multiple pattern searching
(b) Longest common substring detection
(c) Frequent substring identification
(d) Pairwise overlaps computation

(a) Starting point and motivation for the concept of suffix trees was to search for several patterns in a fixed text. Assume we have k patterns of length $O(m)$ each, and a fixed text T of length $O(n)$. Building the suffix tree of T in time $O(n)$ and searching for each pattern as a substring of T, that is a prefix of a suffix of T, we may simply navigate along each pattern into the suffix tree and find all occurrences of the pattern. For each pattern, navigation into the suffix tree requires $O(m)$ steps. Thus the overall complexity of multiple pattern searching is $O(n + km)$, which considerably improves, for example, the Knuth-Morris-Pratt approach that requires $O(k(n + m))$ steps.

(b) Concerning the detection of a *longest common substring* within two strings S and T of length $O(n)$ each, a naive algorithm obviously requires $O(n^3)$ steps. Knuth conjectured that no linear time algorithm would be possible [44]. Using suffix trees this can be done in $O(n)$ steps. Again, note here the surely realistic (concerning sizes of problems in bioinformatics) assumption made with Ukkonen's results that all used numbers can be processed in constant time. Using logarithmic instead of uniform measure changes complexity estimation (and was eventually understood this way in Knuth's conjecture). The proceeding is as follows. First construct a common suffix tree for $S\$_1$ and $T\$_2$ in time $O(n)$. We use two different end markers in order to prevent a common end marker to count for a longest common substring. Then look for nodes labelled 'ST'. Note that below any such node there is always at least one leaf presenting a suffix of $T\$_1$, and at least one leaf representing a suffix of $S\$_2$. Thus any such node has as path label a string that is both a substring of S and a substring of T. Now simply look for a node labelled 'ST' with longest path label. To find such a node requires a further linear time traversal through the suffix tree. Thus the overall complexity is $O(n)$.

(c) Looking for all substrings with at least α and at most β occurrences in a text T of length n is an interesting problem. By choosing an appropriate value for α we can exclude substrings that occur too seldom to be characteristic for text T, and by choosing an appropriate value for β we can exclude trivial strings like 'and' that occur in great number within any long enough text. Though there might be $O(n^2)$ substrings within T, application of the suffix tree for T gives a compact representation of all desired strings which allows a linear time solution. This representation is some sort of interval representation. By this we mean that for a decomposition of a substring into XY such that, for all proper prefixes Y' of Y, substring XY' has at least α and at most β occurrences, whereas XY has less than α occurrences, we may represent the whole ensemble of substrings XY' for all proper prefixes Y' of Y by the "string interval" between X (included) and XY (excluded). What we do with the suffix tree of T is the following. In a bottom-up traversal we compute for every node u the number $f(u)$ of leaves below this node. In a preorder traversal we determine nodes u closest to the root such that $f(u) \geq \alpha$ and below such a node u for all nodes v closest to the root with $f(v) > \beta$. Let w be the predecessor node of node u. Then the string interval between the path label of node w and the path label of node v defines a whole ensemble of substrings with at least α and at most β occurrences in a text T.

(d) Having k strings of length $O(n)$ each, the computations of pairwise overlaps, that is of the overlap matrix required for the sequencing of DNA strings by the shotgun procedure, requires in a naive approach $O(k^2 n^2)$ steps. Using suffix trees this may be considerable improved to $O(kn + k^2)$. The procedure is as follows. Given k strings $S_1, S_2, \ldots, S_k$ of length $O(n)$ each, such that no string occurs as a substring of another one, first construct the common suffix tree of $S_1, S_2, \ldots, S_k$ in time $O(kn)$. Given two strings S_i and S_j, note that the overlap O_{ij} between S_i and S_j can be recovered from the common suffix tree as follows (Fig. 4.62).

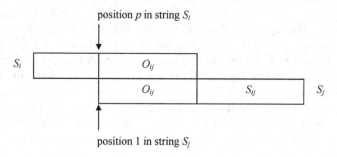

Fig. 4.62. How to detect overlaps in a common suffix tree

String O_{ij} is a suffix of S_i starting at position p, whereas string O_{ij} is a prefix of a suffix of S_j starting at position 1. Navigating to the position with path label O_{ij} in the common suffix tree we arrive at a node u (and not at a position between two consecutive characters of an edge label). The reason is that S_j is not a substring of S_i, thus in string S_j the considered prefix O_{ij} is followed by a character different from $, whereas in string S_i the considered substring O_{ij} is followed by end marker $. Thus there must be a node u at the position with path label O_{ij} splitting into an edge labelled $ and leading to a leaf labelled 'S_j 1', and into a path leading to a leaf labelled 'S_i p' (Fig. 4.63).

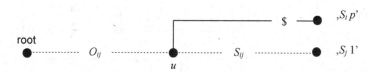

Fig. 4.63. How to detect overlaps in a common suffix tree (continued)

Thus leaf labelled 'S_j 1' tells us that we can read out, along the path π_j between the root and that leaf, all overlaps between any string S_i (note that $i = j$ is allowed) and S_j. Reading out overlaps with S_j on the detected path π_j requires detection of nodes u with a direct $-link to a leaf labelled 'S_i'. Also note that there may exist, for a fixed index i, several such $-links leading to a leaf labelled 'S_i'. The reason is that there may be different suffix-prefix matches between S_i and S_j. Being interested in the overlap, that is, the longest suffix-prefix match between S_i and S_j we must take the deepest node u with an edge labelled '$' leading to a leaf labelled 'S_i'. Summarizing, there are two events that are of interest:

(1) Arriving at a leaf labelled 'S_j 1' for some j.
(2) Finding a deepest node on the path to the leaf found in (1) with edge labelled '$' and leading to a leaf labelled 'S_i p', for some i, p.

Simply waiting for events (1) and, for each such, going back to the root and detecting on the path from the root down to the found leaf all events (2) is not the most efficient implementation extracting all overlaps. The reason is that we visit for each of the n leaves labelled '1' a complete path from the root to each such leaf, leading in the worst case to an $O(k^2)$ running time as different such paths may be visited (at least in parts) in a multiple manner. A better idea is to manage for each $i = 1, \ldots, k$ a separate stack that stores during a preorder traversal through the common suffix tree all nodes u so far found to have a $-link to a leaf labelled 'S_i'. Whenever a leaf labelled 'S_j 1' is found, all top-elements of stacks allow to read out the overlaps of strings with S_j.

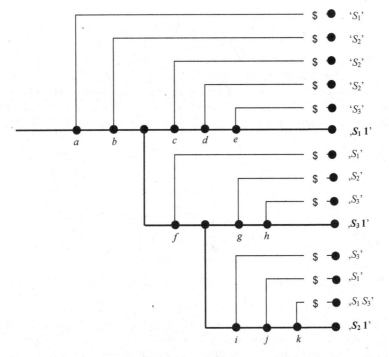

Fig. 4.64. Traversal through common suffix tree to compute all overlaps

Backtracking through the tree leads to the deletion of top-elements of the stacks whenever corresponding nodes are passed, whereas forth-tracking again fills the stacks. We illustrate the evolution of stacks and the read-out of overlaps with strings S_1, S_2, S_3. Assume that parts of a common suffix tree are shown in Fig. 4.64. The evolution of stacks (*forth* going one node deeper, *back* going one node back) and the read-out of overlaps during preorder traversal are as follows (Figs. 4.65 and 4.66). Instead of writing down an overlap in the overlap matrix we indicate it by the node having the overlap as path label.

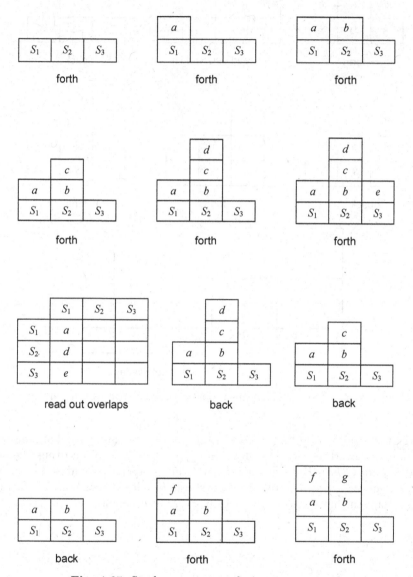

Fig. 4.65. Stack management during tree traversal

Now $O(kn)$ is the complexity of constructing the common suffix tree, another $O(kn)$ is for a preorder traversal through the common suffix tree, $O(k^2)$ is for reading out k^2 overlaps as top-elements of the stacks at particular stages

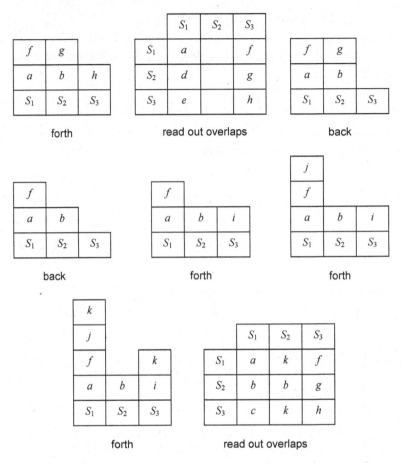

Fig. 4.66. Stack management during tree traversal (continued)

of the traversal. Term $O(k^2)$ deserves some explanation: one could argue that reading out overlaps happens during the preorder traversal and thus should also require $O(kn)$. However, note that it may happen that at some leaves several entries are pushed onto various stacks (see leaf below node k), and it might happen that the same node defines various different overlaps (see node b).

4.4 Least Common Ancestor

4.4.1 Motivation

Assume an operation f is to be executed on a data structure D rather frequently. Then it might be advantageous to preprocess D in advance yielding a data structure D^* on which execution of f is much less expensive than on D. This could even be an advantage if the transformation of D into D^* is more expensive than a single execution of f. A situation like this occurs for the operation of determining the least common ancestor of two nodes in a tree. Having a tree D with n nodes, starting at nodes x and y we may find the least common ancestor z of x and y in D in linear time $3n$ as follows. If $x = y$ then $z = x$. If $x \neq y$ proceed as follows: starting at x, walk towards the root r thereby collecting all visited nodes in a stack $S(x)$. Starting at y, walk towards the root thereby collecting all visited nodes in a stack $S(y)$. Successively delete from $S(x)$ and $S(y)$ identical top elements until for the first time both lists have different top elements. The last deleted common top element is the least common ancestor of x and y. For special trees, e.g. balanced trees, execution time may be even better, e.g. $3 \log n$. With a suitable preprocessing of D that requires linear time we may even arrive at a constant execution time c for every least common ancestor computation, regardless of how many nodes the tree has (shown in [34]; see also [67]). Expecting a large number k of computations of least common ancestors, even linear preprocessing costs may be advantageous. Such possible advantages of preprocessing are well known from other applications, e.g. for searching in arrays. Assume we expect k lookups for elements in an array A of length n. Without any preprocessing this can be done in kn steps. After sorting the array in time $O(n \log n)$ the same task can be done in $O(k \log n)$ steps.

4.4.2 Least Common Ancestor in Full Binary Trees

As a preparation for the case of arbitrary trees we treat the problem of computing least common ancestors in full binary trees. The problem will be much easier solved than for the general case, nevertheless it shows the general principle and the notions introduced here will be required later, too. So let a full binary tree with n nodes be given. First do an inorder traversal through the tree D:

- For a tree consisting of only one node, simply visit this node.
- For a tree consisting of more than a single node, recursively traverse its left subtree, then visit its root, finally recursively traverse its right subtree.

Attach to each node its inorder number written as a binary number with d bits (leading zeroes are admissible). In this section we always refer to nodes by their inorder numbers. In the example shown in Fig. 4.67 we obtain the following inorder numbers (binary code written inside the boxes that represent nodes).

Particular useful information is contained within a certain prefix of inorder numbers (in the diagram in Fig. 4.67 indicated by bold face prefix of each number). This is defined using the decomposition of every non-zero binary number z having d bits into $z = \pi(z)10^{\text{right}(z)}$ with a suffix block of right(z) many zeroes, preceded by rightmost bit 1, and preceded by a prefix string $\pi(z)$. Later we will also make use of the similar composition $z = 0^{\text{left}(z)}1\sigma(z)$ into a prefix block of left(z) many zeroes, followed by leftmost bit 1, followed by remaining suffix string $\sigma(z)$.

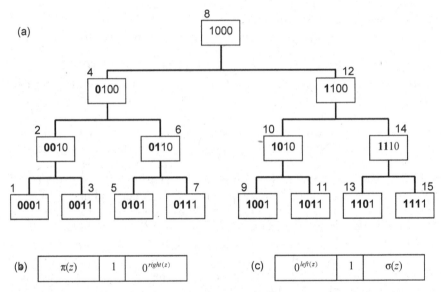

Fig. 4.67. (a) inorder traversal and inorder numbers; (b) decomposition according to rightmost 1; (c) decomposition according to leftmost 1

Prefix string $\pi(x)$ always encodes the path from the root of D to the node z in the usual sense that bit 0 encodes going to the left subtree and bit 1 going to the right subtree. The reader may check this for the example above, and also convince himself that this statement is true for an arbitrary binary tree (argue recursively). We require a second numbering of the nodes, namely the one resulting from a preorder traversal through the tree:

- For a tree consisting of only one node, simply visit this node.
- For a tree consisting of more than a single node, first visit its root, then recursively traverse its left subtree, finally recursively traverse its right subtree.

Attach to each node x binary number pre(x) that corresponds to the step at which node x is visited in a preorder traversal; call pre(x) the *preorder number* of x. In Fig. 4.68 we obtain the following preorder numbers (written inside the node boxes). Finally, we make a third traversal through the tree and attach to each node x the number $f(x)$ of nodes in the subtree with root x. To equip each node with these numbers requires $O(n)$ steps (three tree traversals).

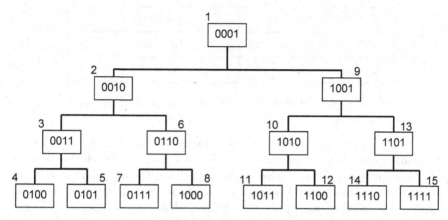

Fig. 4.68. Preorder traversal and preorder numbers

Having available these numbers we may now compute the least common ancestor z (strictly speaking, its inorder number) for arbitrary nodes x and y in constant time, regardless of how great the number of nodes in the tree is. Using $x = \pi(x)10^{\text{right}(x)}$ and $y = \pi(y)10^{\text{right}(y)}$, we have to compute $z = \pi(z)10^{\text{right}(z)}$. Since $\pi(x)$ and $\pi(y)$ encode the path from the root down to x and y, respectively, we conclude that their longest common prefix encodes the path from the root down to the least common ancestor z of x and y. Thus the longest common prefix of strings $\pi(x)$ and $\pi(y)$ equals $\pi(z)$. For the moment, denote the length of $\pi(z)$ by q. The computation of z now proceeds as follows:

- If $x = y$ then $z = x = y$ is returned.
- If $x \neq y$ and $\text{pre}(x) < \text{pre}(y) < \text{pre}(x) + f(x)$ then x is a predecessor[2] of y, thus $z = x$ and $z = x$ is returned.
- If $x \neq y$ and $\text{pre}(y) < \text{pre}(x) < \text{pre}(y) + f(y)$ then y is a predecessor of x, thus $z = y$ and $z = y$ is returned.
- In any other case we know that neither x is an ancestor of y, nor is y an ancestor of x. This means that common prefix p of $\pi(x)$ and $\pi(y)$ is strictly shorter than each of $\pi(x)$ and $\pi(y)$. Now we compute z as shown in Fig. 4.69. Here, we use standard bit string operations: arithmetic operations; Boolean operations, shifting a bit string a predefined number of bits to the left or to the right; determining position of left-most resp. right-most bit 1. Availablilty of these operations on a computer and execution time are discussed below.

---

[2] The preorder numbers of all nodes in the subtree with root x are $\text{pre}(x)$, $\text{pre}(x) + 1, \ldots, \text{pre}(x) + f(x) - 1$.

compute	result returned
x XOR y	$0^q 1w$ (for some string w)
$left(0^q 1w)$	q
$shift - right(x, d - q)$	$0^{d-q} \pi(z)$
$0^{d-q} \pi(z) + 0^{d-q} \pi(z) + 0^{d-1} 1$	$0^{d-q-1} \pi(z) 1$
$shift - left(0^{d-q-1} \pi(z) 1, d - q - 1)$	$\pi(z) 1 0^{d-q-1}$

Fig. 4.69. Computation of least common ancestor

In this computation we make the assumption that the length $d = O(\log n)$ of the binary numbers under consideration is small enough[3] such that certain bit string operations are available on the underlying machine and can thus be executed in constant time. Usually, on a machine arithmetical operations as addition, subtraction, multiplication, division, and comparison are available, as are logical operations like AND, OR and XOR. Operation shift-left(z, i) can be simulated by multiplying z with 2^i, shift-right(z, i) can be simulated by division of z by 2^i. Numbers 2^i can be created by an operation setting a single bit 1 at a specified position.

The computation of operations left(z) and right(z) is not necessarily present in an arbitrary programming language. So we treat these operations separately in more detail and show how to make them being computable in constant time. First we consider the operation right(z) computing the number of zeroes at the right end of z. Instead of trying to simulate right(z) by other available operations executable in constant time, we simply make a further preprocessing step (in linear time) that provides us with all the information necessary for the computation of right(z), for all binary numbers from 1 to 2^d. First we walk through all powers of 2 (successively multiplying with 2) and always increase right(z) by 1. This gives the correct numbers right(z) for all powers of 2. For all numbers z between 2^i and 2^{i+1} we obtain right(z) in a second walk by simply copying all already computed numbers right$(1), \ldots,$ right$(2^i - 1)$. The complete procedure requires linear time. The computation of left(z) for all numbers $z = 1, \ldots, 2^d$ is even simpler: For powers of 2, left(1) begins with $d - 1$, and decreases one by one down to left$(2^d) = 0$. For all numbers z between 2^i and $2^{i+1} - 1$, left(z) equals left$(2i)$. Figure 4.70 shows computations for $d = 4$.

---

[3] For example, length $d = 64$ allows consideration of trees up to 2^{64} nodes. This is much more than is ever needed in practical applications.

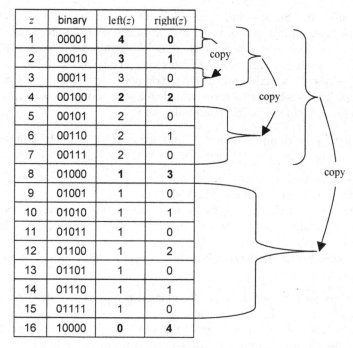

z	binary	left(z)	right(z)
1	00001	4	0
2	00010	3	1
3	00011	3	0
4	00100	2	2
5	00101	2	0
6	00110	2	1
7	00111	2	0
8	01000	1	3
9	01001	1	0
10	01010	1	1
11	01011	1	0
12	01100	1	2
13	01101	1	0
14	01110	1	1
15	01111	1	0
16	10000	0	4

Fig. 4.70. Computation of values left(z) and right(z), for all d-bit numbers

4.4.3 Least Common Ancestor in Arbitrary Trees

Now we generalize the ideas above to the case of arbitrary trees. Two difficulties must be overcome:

- Nodes may have more than two successor nodes.
- Binary trees must not be full binary trees.

Whereas preorder numbers are still defined in arbitrary trees, inorder numbers make sense only for binary trees. We make again use of the preorder numbers of the nodes of an arbitrary tree with n nodes. For simplicity of presentation, in this section we identify nodes with their preorder numbers. Again we assume that all these numbers are of the same length d and achieve this by eventually adding leading zeroes. Note that the decomposition $x = \pi(x)10^{\text{right}(x)}$ introduced for arbitrary binary strings in the section before is applicable also for preorder numbers.

Lemma 4.6.

For every node u, the subtree with root u contains exactly one node v having maximum value right(v) *among all nodes within subtree below u. Call this unique node $I(u)$.*

Proof. Assume, the subtree with root u contains two different nodes v and w having maximum value right(v) = right(w) = i. Hence, both v and w have the common suffix 10^i. Since v and w are different nodes there is a position in v and w with different bits. We consider the leftmost such position. Thus we may decompose, without loss of generality, v and w as follows: $v = Z1V10^i$ and $w = Z0W10^i$ with common prefix Z. Now consider node $z = Z100\ldots0$ with sufficient zeroes at the right end to give a string of length d. Thus $w < z < v$ is true. Hence, z also is located within the subtree with root u. However, $\text{right}(z)$ is greater than i, contradicting maximality of i.   □

For each node u we compute node $I(u)$ and attach it to node u by a bottom-up traversal through the tree in linear time as follows.

- For every leaf u, obviously $I(u) = u$.
- For every inner node u we distinguish two cases.
  (a) If $\text{right}(u) \leq \text{right}(I(v))$ for at least one child node v of u, choose the unique child node v of u with maximum value $\text{right}(I(v))$ and define $I(u) = I(v)$.
  (b) If $\text{right}(u) > \text{right}(I(v))$ holds for every child node v of u, then $I(u) = u$.

Since every inner node u has at most one child v with $I(u) = I(v)$, the nodes u with identical value $I(u)$ form a path in the tree, called the *component* of u. The deepest node of the component of u is obviously $I(u)$, its highest node is called the *head* of the component. When computing the values $I(u)$ for every node u, we assume that a *pointer* from the deepest node of each component to its head is provided. This is simply achieved by starting with a pointer for each leaf pointing to the corresponding leaf. Then the pointer is lifted to the parent node in case (a) of the computation of $I(u)$, whereas in case (b) node u receives a fresh link pointing to himself (Fig. 4.71).

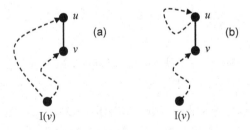

Fig. 4.71. Lifting of pointers versus creation of a new pointer

Finally, some further information is required at each node u, namely the following bit string

$$A_u = A_u(d-1)\ldots A_u(1)A_u(0) \tag{4.4}$$

with i^{th} bit defined as follows: $A_u(i) = 1$ if there is an ancestor v of u (eventually u itself) with $\text{right}(I(v)) = i$, and $A_u(i) = 0$ if there is no ancestor v

of u with right$(I(v)) = i$. By a further preorder traversal through the tree we may easily attach bit string A_u to each node u:

- If u is the root then $A_u = 0^{i-1}1^{d-i-1}$ with $i = \text{right}(I(u))$.
- If u is not the root take its parent v with already computed bit string A_v. Then we obtain $A_u = A_v$ OR $0^{i-1}1^{d-i-1}$ with $i = \text{right}(I(u))$.

Correctness of the first case is clear, the latter case is correct since ancestors of v are just v itself as well as all ancestors of u.

In Fig. 4.72, components consisting of two or more nodes are indicated by a shadowed area. Each of the nodes outside these areas forms a separate component consisting of this single node. Furthermore, Fig. 4.73 shows all bit strings A_u.

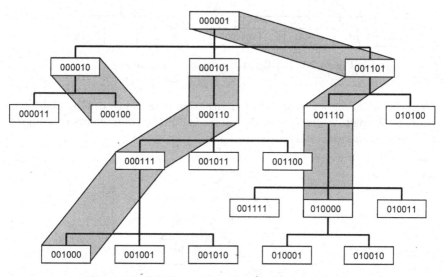

Fig. 4.72. (a) Components having identical I-value

Lemma 4.7. *Representation Lemma*
If x is an ancestor of y, then $\pi(I(x))$ is a prefix of $\pi(I(y))$, and thus right$(I(x)) \geq$ right$(I(y))$. If x is an ancestor of y and $I(x) \neq I(y)$, then $\pi(I(x))$ is a proper prefix of $\pi(I(y))$, and thus right$(I(x)) >$ right$(I(y))$.

The meaning of the *Representation Lemma* is that function I can be seen as an ancestor relationship conserving embedding of an arbitrary tree into a full binary tree. After application of I, ancestor information for nodes x and y can be extracted almost the same way as it was done for full binary trees.

Proof. If x and y are nodes within the same component then $I(x) = I(y)$ and the assertion of the lemma is trivial. So assume x and y are located

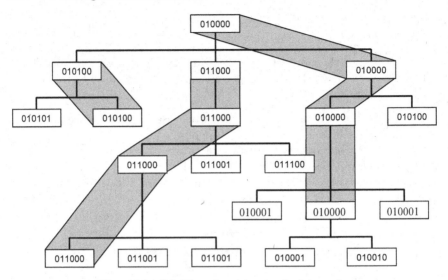

Fig. 4.73. (b) Bit strings A_u attached to nodes

within different components. Thus the situation is as shown in Fig. 4.74 (only segments between x and $I(x)$, and y and $I(y)$ of the components of x and y are drawn as shaded areas).

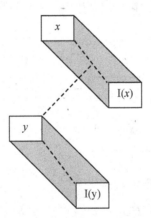

Fig. 4.74. Different components of x and y

Decompose $I(x) = \pi(I(x))10^{\text{right}(I(x))}$ and $I(y) = \pi(I(y))10^{\text{right}(I(y))}$. Since x and y are located in different components we know that $\text{right}(I(x)) > \text{right}(I(y))$. Now assume that $\pi(I(x))$ is not a prefix of $\pi(I(y))$. Let P be longest common prefix of $\pi(I(x))$ and $\pi(I(y))$, and let p be its length. Thus P is followed by different bits in $I(x)$ and $I(y)$. Consider $w = P10^{d-p-1}$. If $I(x) = P1U10^{\text{right}(I(x))}$ and $I(y) = P0V10^{\text{right}(I(y))}$ then $I(y) < w < I(x)$

holds. If $I(x) = P0U10^{\text{right}(I(x))}$ and $I(y) = P1V10^{\text{right}(I(y))}$ then $I(x) < w < I(y)$ holds. As $I(x)$ and $I(y)$ are both located within the subtree below node x, we conclude for both cases that w is located within the subtree below node x, too. Furthermore $\text{right}(w) > \text{right}(I(x))$. This contradicts maximality of $\text{right}(I(x))$. □

Lemma 4.8.
For nodes x and y let z be their least common ancestor. Then $k = \text{right}(I(z))$ can be computed from x and y in constant time having available all the information provided so far at each node of the considered tree, as well as the information necessary to do all the operations discussed in the section before for a full binary tree with inorder numbers of bit length d.

Proof. Since z is an ancestor of both x and y, *Representation Lemma* tells us that decompositions of $I(x)$, $I(y)$, and $I(z)$ must be related as follows, with strings P, X, and Y:

$$I(z) = \pi(I(z))10^{\text{right}(I(z))}$$
$$I(x) = \pi(I(x))10^{\text{right}(I(x))} = \pi(I(z))PX10^{\text{right}(I(x))}$$
$$I(y) = \pi(I(y))10^{\text{right}(I(y))} = \pi(I(z))PY10^{\text{right}(I(y))} \ .$$

Thus P is the maximum common substring of $\pi(I(x))$ and $\pi(I(y))$ starting right of $\pi(I(z))$. Note that one or both of strings X, Y may be empty (for example in case that $I(x) = I(y)$). In case that both X, Y are non-empty strings X, Y have different leftmost characters. From preorder numbers $I(x)$ and $I(y)$ available at nodes x and y we compute as done in Sect. 4.4.2 string g defined as follows:

$$g = \pi(I(z))P10^{\text{right}(g)}$$
(g and $\text{right}(g)$ are computed in constant time) .

Comparing decompositions of $I(z)$ and g we infer that $\text{right}(g) \leq \text{right}(I(z))$ holds. We know that z is an ancestor of x, therefore the definition of string A_x tells us that $A_x(\text{right}(I(z))) = 1$. By the same way, as z is also an ancestor of y, definition of string A_y tells us that $A_y(\text{right}(I(z))) = 1$. So far, we have shown that $\text{right}(I(z))$ (the number that is to be computed) fulfils the following constraints:

$$\text{right}(I(z)) \geq \text{right}(g)$$
$$A_x(\text{right}(I(z))) = 1$$
$$A_y(\text{right}(I(z))) = 1 \ .$$

At this stage consider the least number k that fulfils these constraints:

$$k \geq \text{right}(g)$$
$$A_x(k) = 1$$
$$A_y(k) = 1 \ .$$

Remembering the numbering of indices of A_x and A_y from left to right, namely from $d-1$ down to 0 as shown in Fig. 4.75, and using abbreviation $m = right(g)$, number k is computed in constant time using available bit strings A_x, A_y, g and number $m = right(g)$ as shown in Fig. 4.76.

	$d-1$	...	k	...	$m-1$	...	0
A_x	...	...	1	0 0 ... 0 0	...	...	...
A_y	...	...	1	0 0 ... 0 0	...	...	...

search here for right-most common 1

Fig. 4.75. Where to search for k

compute	result returned
$shift-right(A_x, m-1)$	U
$shift-right(A_y, m-1)$	V
U AND V	W
$right(W)$	$k-m$
add m	k

Fig. 4.76. Computation of $k = right(I(z))$

Finally we show that number k just computed and desired number $right(I(z))$ coincide. Since numbers k and $right(I(z))$ share the three constraints above and k was defined as the least such number we conclude that $k \leq right(I(z))$. Conversely, we again interpret the definitions of $A_x(k) = A_y(k) = 1$ and obtain that node x has an ancestor u such that $right(I(u)) = k$, and node y has an ancestor v such that $right(I(v)) = k$. A further application of the *Representation Lemma* allows us to infer decompositions of $I(u)$ and $I(v)$ from the formerly derived decompositions of $I(x)$ and $I(y)$.

$$I(x) = \pi(I(x))10^{right(I(x))} = \pi(I(z))PX10^{right(I(x))}$$
$$I(y) = \pi(I(y))10^{right(I(y))} = \pi(I(z))PY10^{right(I(y))}$$
$$I(u) = \pi(I(u))10^k \text{ with } \pi(I(u)) \text{ being a prefix of } \pi(I(z))PX$$
$$I(v) = \pi(I(v))10^k \text{ with } \pi(I(v)) \text{ being a prefix of } \pi(I(z))PY$$

From $k \geq \text{right}(g)$ and the decomposition $g = \pi(I(z))P10^{\text{right}(g)}$ we conclude that $\pi(I(u))$ and $\pi(I(v))$ do not have greater length than $\pi(I(z))P$ has. As $\pi(I(u))$ is a prefix of $\pi(I(z))PX$ and $\pi(I(v))$ is a prefix of $\pi(I(z))PY$, we conclude that $\pi(I(u)) = \pi(I(v))$, thus $I(u) = I(v)$. This means that nodes u and v are located within the same component. This means that one of u, v is an ancestor of the other one. Assume that u is an ancestor of v. Since u is also an ancestor of x and v is an ancestor of y we know that u is a common ancestor of x and y. Thus u is an ancestor of the least common ancestor z of x and y. A final application of the *Representation Lemma* tells us that $\pi(I(u))$ is a prefix of $\pi(I(z))$. A further look at the decompositions $I(u) = \pi(I(u))10^k$ and $I(z) = \pi(I(z))10^{\text{right}(I(z))}$ then yields $k \geq \text{right}(I(z))$, the converse estimation.   □

Lemma 4.9.
Given nodes x and y, their least common ancestor z can be computed in constant time.

Proof. Given x and y with least common ancestor z and using Lemma 4.8, we first compute $\text{right}(I(z))$. Now we consider the component of z. Since z is a predecessor of x and y we conclude from the *Representation Lemma* that $\text{right}(I(z)) \geq \text{right}(I(x))$ and $\text{right}(I(z)) \geq \text{right}(I(y))$. Now we walk down along the path from z to x and also walk down along the path from z to y until when we eventually leave the component of z. If we do so we look for the first node outside the component of z. Several cases must be distinguished.

(1) We did not leave the component of z, thus $I(x) = I(y) = I(z)$ and $\text{right}(I(x)) = \text{right}(I(y)) = \text{right}(I(z))$. So, either x is an ancestor of y (Fig. 4.77 (1)), or $x = y$, or y is an ancestor of x. In any case, z is obtained as the smaller one of x and y.

(2) Only the path from z down to y leaves the component of z. In this case $\text{right}(I(x)) = \text{right}(I(z))$ and $\text{right}(I(z)) > \text{right}(I(y))$. Let w be the first node outside the component of z on the path from z down to y (Fig. 4.77 (2)). We show below how to compute w in constant time. From w we obtain z as the immediate predecessor of w.

(3) A symmetric situation as in (2) is that only the path from z down to x leaves the component of z. In this case $\text{right}(I(y)) = \text{right}(I(z))$ and $\text{right}(I(z)) > \text{right}(I(x))$. As in (2) let w be the first node outside the component of z on the path from z down to x. We show below how to compute w in constant time. From w we obtain z as the immediate predecessor of w.

(4) Both paths from z down to x and from z down to y leave the component of z. In this case $\text{right}(I(z)) > \text{right}(I(x))$ and $\text{right}(I(z)) > \text{right}(I(y))$. Let v be the first node outside the component of z on the path from z down to x, and w be the first node outside the component of z on the path from z down to y. We show below how to compute v and w in constant time. From v and w we obtain z as the smaller one of the immediate

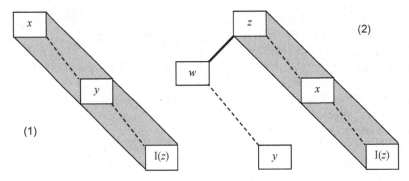

Fig. 4.77. (1) x, y within component of z; (2) x within, y outside component of z

predecessors of v and w (in Fig. 4.78, the smaller one of the predecessors is the predecessor of w).

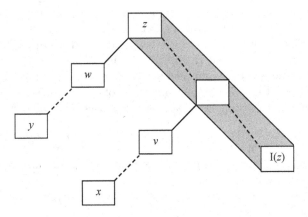

Fig. 4.78. x, y outside component of z

First note that the distinction into these four cases can be computed in constant time as follows:

- $\mathrm{right}(I(z)) = \mathrm{right}(I(x))$ and $\mathrm{right}(I(z)) = \mathrm{right}(I(y))$ for case (1)
- $\mathrm{right}(I(z)) = \mathrm{right}(I(x))$ and $\mathrm{right}(I(z)) > \mathrm{right}(I(y))$ for case (2)
- $\mathrm{right}(I(z)) > \mathrm{right}(I(x))$ and $\mathrm{right}(I(z)) = \mathrm{right}(I(y))$ for case (3)
- $\mathrm{right}(I(z)) > \mathrm{right}(I(x))$ and $\mathrm{right}(I(z)) > \mathrm{right}(I(y))$ for case (4)

Next we consider node v, for example in case (4), and first show how to compute number $\mathrm{right}(I(v))$ in constant time. The treatment of node v in cases (2) and (3), and of node w in case (4) is done exactly the same way. Since v is an ancestor of x, the values $\mathrm{right}(I(\ldots))$ occurring along the path from z down to x are constant within components and decreasing when leaving a component. As v is the first node on that path leaving the component of z we know that $\text{right}(I(v))$ can be found in A_x as the greatest index $j <$ $\text{right}(I(z)) = k$ with $A_x(j) = 1$ (Fig. 4.79). This unique index j can thus be computed in constant time as shown in Fig. 4.80.

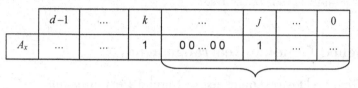

Fig. 4.79. Where to search for j

compute	result returned
$shift-left(A_x, d-k)$	W
$left(W)$	$j-k$
add k	j

Fig. 4.80. Compute j

Having computed number $j = \text{right}(I(v))$, we finally compute nodes $I(v)$ and v. To see how this can be done make a final use of decomposition $I(v) = \pi(I(v))10^{\text{right}(I(v))}$. Since v is an ancestor of x we know from the *Representation Lemma* that $\pi(I(v))$ is a prefix of $\pi(I(x))$, in particular a prefix of $I(x)$. Knowing $j = \text{right}(I(x))$ we may thus extract $\pi(I(x))$ as prefix of $I(x)$ of length $d-1-j$ and compute $I(v)$ as in Fig. 4.81. Finally compute v

compute	result returned
$shift-right(I(x), j+1)$	$0^{j+1}\pi(I(x))$
$0^{j+1}\pi(I(x)) + 0^{j+1}\pi(I(x)) + 0^{d-1}1$	$0^{j}\pi(I(x))1$
$shift-left(0^{j}\pi(I(x))1, j)$	$\pi(I(x))10^{j} = I(v)$

Fig. 4.81. Compute $I(v)$

as the head of its component by following the extra pointer from $I(v)$ to the head of the component of v. □

Exercise

4.1. Verify bit strings A_u computed for the example in Fig. 4.73. Why is A_u identical for all nodes u within one component? Why is the number of bits 1 in A_u identical to the number of components that are visited on the path between u and the root of the tree?

4.5 Signed Genome Rearrangement

4.5.1 Reality-Desire Diagrams of Signed Permutations

A *signed permutation* of n numbers is a sequence $\delta_1\pi(1)\delta_2\pi(2)\ldots\delta_n\pi(n)$ with a permutation π of numbers $1, 2, \ldots, n$, and algebraic signs (+ or -) $\delta_1, \delta_2, \ldots, \delta_n$. To execute a *reversal* between positions i and j of a signed permutation means to invert order and change signs of all entries between entry i and entry j. *Sorting* a signed permutation means transforming it by a sequence of reversals into the *sorted permutation* $+1 + 2 \ldots + n$. The minimum number of reversals required to sort a signed permutation is called its *reversal distance*. In Setubal & Meidanis ([68]) a visualization of signed permutations is proposed that turns out to be particularly helpful in the explanation of the working of the algorithm of Hannenhalli & Pevzner for the determination of reversal distance. It is the so-called *Reality-Desire diagram (RD-diagram)* of a signed permutation $\delta_1\pi(1)\delta_2\pi(2)\ldots\delta_n\pi(n)$. In a RD-diagram signed number $+g$ is represented by a succession of two nodes $+g, -g$, whereas signed number $-g$ is represented by the reversed succession of nodes $-g, +g$. For signed number δg let $-\delta g$ be the same number, but with reversed sign.

Using two additional nodes L and R we arrange counter-clockwise around a circle in succession node pair L, R and all node pairs corresponding to signed numbers $\delta_1\pi(1)\delta_2\pi(2)\ldots\delta_n\pi(n)$. Starting with node L and proceeding counter-clockwise we connect every second node with its successor node by *reality links (R-links)*: L with $\delta_1\pi(1)$, $-\delta_1\pi(1)$ with $\delta_2\pi(2)$, ..., $-\delta_{n-1}\pi(n-1)$ with $\delta_n\pi(n)$, $-\delta_n\pi(n)$ with R. Reality links are thus arranged along the borderline of the circle according to the order of signed numbers as it is at the beginning ("reality"). Similarly, *desire links (D-links)* are introduced to reflect the ordering $+1 + 2 \ldots + n$ of genes as it should be in the end ("desire"). We connect L with $+1$, -1 with $+2$, ..., $-(n-1)$ with $+n$, $-n$ with R. The resulting diagram with $2n + 2$ nodes, $n + 1$ R-links and $n + 1$ D-links is called the *RD-diagram* of the signed permutation under consideration. Figures 4.82 and 4.83 present examples to clarify the construction. For better visibility, we sometimes draw D-links as curves (though they are, strictly speaking, lines; being lines will indeed be important in later constructions).

After having introduced R-links and D-links into the RD-diagram, the labeling of nodes, with the exception of nodes L and R, does not play a role any more, thus we omit it in all of the following examples. Take a look at

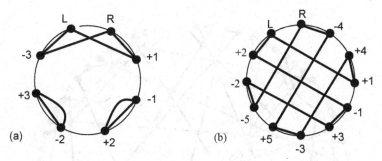

Fig. 4.82. Signed permutations: **(a)** -3 -2 -1; **(b)** +2 -5 -3 -1 +4

a more complex example with 36 signed numbers. Its RD-diagram is shown in Fig. 4.83. As within the example diagrams above, we again observe that R-links and D-links form *alternating cycles* consisting of an alternation of R-links and D-links. This is true in general for the following reason: each node has exactly one adjacent R-link and one adjacent D-link. So, following links always proceeds in an alternation of R-links and D-links, and must end up with the start node. In Fig. 4.83 try to detect all 11 cycles[4].

In the following it is important to have a clear imagination of how reversals affect RD-diagrams. First, the *reversed area* corresponds to a segment of the RD-diagram between any two R-links. Note that R-links correspond to the cutting points between consecutive genes. Of course, the segment that corresponds to the reversed area may not contain nodes L and R. Reversing now means to mirror all nodes in the reversed area (this corresponds to an inversion of the order of genes together with an inversion of signs) and let the D-links of the nodes in the reversed area walk with the nodes to their new position (D-links represent the unchanged target permutation). We illustrate the effect of reversals on RD-diagrams by two examples in Fig. 4.84 (the reversed area is always indicated by a shadowed segment). What can be seen is that both reversals split a cycle into two cycles. This is advantageous having in mind what is finally to be achieved, namely the sorted permutation that corresponds to the trivial RD-diagram shown below having only trivial cycles (consisting of two nodes). Nevertheless, we will later see that reversal (a) has serious disadvantages compared to reversal (b).

What finally results after having sorted the initial permutation is shown in Fig. 4.85.

Unfortunately, not every reversal splits a cycle into two. For example, inverting one of the reversals in Fig. 4.84 has the effect of melting two cycles into a single one. There are also examples of reversals leaving the number of cycles unchanged (Fig. 4.86).

---

[4] A useful tool for displaying RD-diagrams may be found under www.inf.uos.de/theo/public.

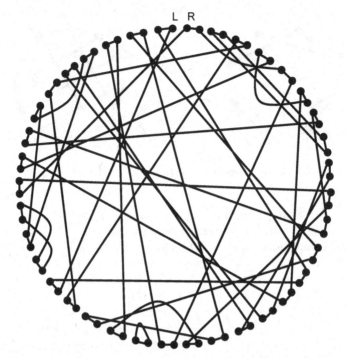

Fig. 4.83. Signed permutation: -12 +31 +34 -28 -26 +17 +29 +4 +9 -36 -18 +35 +19 +1 -16 +14 +32 +33 +22 +15 -11 -27 -5 -20 +13 -30 -23 +10 +6 +3 +24 +21 +8 +25 +2 +7

How a reversal affects the number of cycles can be easily characterized. Given two different R-links within the same cycle, we call them *convergent* in case thát in a traversal of the cycle both links are traversed clockwise or both links are traversed counter-clockwise (as links occuring on the circle defining the RD-diagram). Otherwise we call them *divergent*. Note that in Fig. 4.84, (a) and (b) the reversed area was between two divergent R-links, whereas in Fig. 4.86 it was between two convergent R-links.

Lemma 4.10.
A reversal changes number of cycles by +1 in case it acts on divergent R-links of the same cycle, by 0 in case it acts on convergent R-links of the same cycle, and by −1 in case it acts on R-links of different cycles.

Proof. The generic diagrams in Fig. 4.87 make clear what happens to cycles after reversals. Broken lines always indicate paths (that may, of course, be much more complicated that the diagrams suggest; nevertheless, arguments can be made clear even with these rather simple drawings). Figure 4.87 contains reversals acting on (a) divergent R-links of a single cycle; (b) convergent R-links of a single cycle; (c) R-links of two cycles.   □

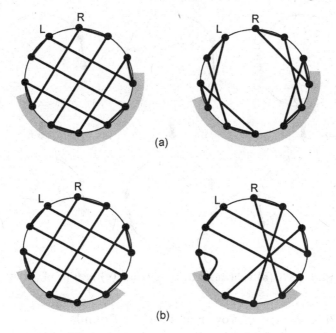

(a)

(b)

Fig. 4.84. Reversals splitting a cycle

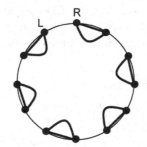

Fig. 4.85. RD-diagram after sorting

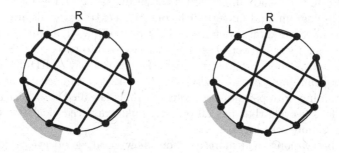

Fig. 4.86. Reversal that leaves number of cycles unchanged

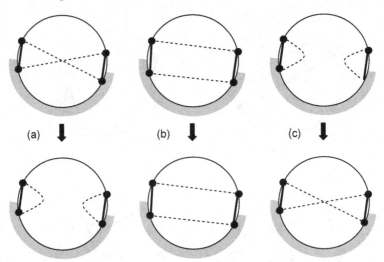

Fig. 4.87. (a) splitting; (b) conserving; (c) melting of cycles

4.5.2 ''Ping-Pong'' of Lower and Upper Bounds

The lemma above leads to a first lower bound for the reversal distance d of a signed permutation of n genes having an RD-diagram with k cycles. Since we start with k cycles and each reversal increases this number by at most 1, we require at least $n + 1 - k$ reversals to achieve the desired RD-diagram having $n + 1$ (trivial) cycles.

Lemma 4.11. *Simple lower bound*

$$n + 1 - k \le d.$$

It rapidly turns out that $n + 1 - k$ is by no means also an upper bound, that is, we cannot always achieve the trivial RD-diagram within $n + 1 - k$ reversals. In the following, we study under which conditions $n + 1 - k$ reversals suffice to sort an RD-diagram, and under which conditions extra reversals are required. As long as an RD-diagram contains at least one cycle with a pair of divergent R-links we can apply a reversal splitting this cycle into two cycles. Since such a reversal makes good progress towards the desired trivial RD-diagram, we call a cycle containing a pair of divergent R-links a *good cycle*. A non-trivial cycle containing only pairs of convergent R-links is called a *bad cycle*. Thus, a bad cycle is a cycle that has all of its R-links traversed clockwise or all of its R-links counter-clockwise.

Two observations shed more light on "how good" good cycles and "how bad" bad cycles really are. First, bad cycles not necessarily force us to introduce extra reversals. Second, good cycles with an uncritical choice of divergent R-links may create bad cycles. The latter was already demonstrated with the example in Fig. 4.84 (a). The former can be seen in the example in Fig. 4.88. The bad 4-cycle starting at node R is transformed ("twisted") by the reversal into a good cycle.

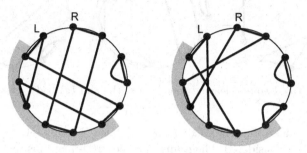

Fig. 4.88. Bad cycle is twisted into a good cycle

Twisting of a bad cycle and thus transforming it into a good cycle happens if and only if the bad cycle has nodes within the reversed area as well as nodes outside the reversed area. This shows that interleaving of cycles will play a central role in sorting RD-diagrams. We fix some notation. Two non-trivial cycles C and C' are *interleaving* if C contains a D-link d and C' a D-link d' such that d and d', drawn as lines, intersect themselves. Interleaving defines a graph structure on the set of non-trivial cycles. Connected components of this graph are simply called *components* of the RD-diagram. A component containing at least one good cycle is itself called a *good component*. A component consisting solely of bad cycles is called a *bad component*. Clearly note the difference between good and bad components: good components contain at least one good cycle, bad components contain only bad cycles. Also note that a component consisting of a trivial cycle is neither called good nor bad. Concerning good components the following result holds; its proof will be postponed to a later section.

Lemma 4.12. *Main Lemma*
Within every good component there exists at least one good cycle with a pair of suitably chosen divergent R-links such that the reversal between these R-links does not produce any bad components.

Remember the example showing that divergent R-links must indeed be chosen very carefully. The following example in Fig. 4.89 (a) gives an indication why proof of the *Main Lemma* is to be expected to be non-trivial. It shows a "diagram" which seemingly is a counter-example to the *Main Lemma*. The "diagram" exclusively consists of good cycles. Consider the good component consisting of two interleaving good cycles. Regardless which of these interleaving cycles we choose, the reversal between its divergent R-links always transforms the other of the two cycles into a bad cycle (Fig. 4.89 (b) and (c)).

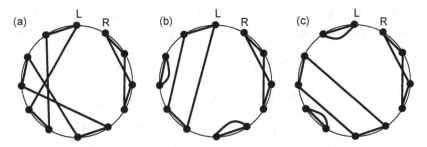

Fig. 4.89. (a) Two interleaving good cycles; (b) and (c) bad cycles generated

Does this mean that the *Main Lemma* is false? Fortunately this is not the case, since the considered "diagrams" are not RD-diagrams as can be seen in Fig. 4.90 by following the D-links, starting at node L. To correctly follow D-links means that after entering some node δg by a D-link, we must next switch to $-\delta g$ (not to the node linked with δg by an R-link) and follow the D-link starting at node $-\delta g$. The resulting D-link traversal is shown in Fig. 4.90 by a sequence of arrows. We observe that not all nodes and D-links have been traversed when ending at node R.

Fig. 4.90. Following D-links, omitting nodes

Thus, in order to prove the *Main Lemma* we must clarify what "diagrams" occur as true RD-diagrams of signed permutations. This is not easily done, but fortunately a rather simple necessary condition for a "diagram" to be a true RD-diagram is available - and suffices to prove the *Main Lemma*.

Lemma 4.13. *Necessary condition of RD-diagrams*
In RD-diagrams, every D-link with convergent adjacent R-links interleaves with at least two D-links (not necessarily belonging to the same cycle).

Proof. We consider a traversal of D-links, starting with link d (Fig. 4.91 (a)). As long as we stay within the region of the cycle above line d and do not come back to start node x, there will always be unvisited nodes (at least one) above line d. Since there are also unvisited nodes below line d, we must at least once traverse a D-link that intersects D-link d, and must again return to the area above d in order to visit the so far unvisited nodes above d. This forces us to traverse a second time across D-link d (Fig. 4.91 (b)).   □

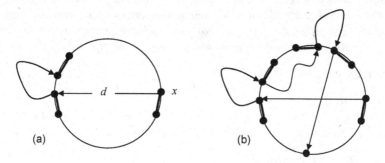

Fig. 4.91. (a) D-link traversal starting; (b) crossing link d

Looking at the "counter-example" in Fig. 4.89 we see that the necessary condition is violated for the diagrams in (b) and (c), thus these cannot be RD-diagrams. Resulting from the diagram in (a) by applying reversals, this diagram cannot be a RD-diagram either.

Iterated application of the *Main Lemma* now allows us to steadily increase by 1 the number of cycles within a good component without running into bad components. Thus we get the following conclusion.

Corollary 4.14.
The cycles of a good component consisting of $2n$ nodes and k cycles can be transformed into trivial cycles by $n - k$ suitably chosen reversals.

Having bad components, the situation is not comparably good. Whatever reversal we apply R-links of a bad cycle (that are thus convergent), the number of cycles does not increase. Thus, we need extra reversals to cope with bad components. Indeed, if we had one extra reversal available for each bad component, an RD-diagram could be sorted as the following statement shows.

Lemma 4.15. *Bad component destruction*
Any reversal acting on two arbitrarily chosen R-links of an arbitrary cycle C in a bad component K turns K into a good component (though cycle C might stay to be a bad cycle).

For the proof as well as for later use we require further notions concerning cycles that are affected by a reversal (in that either their quality changes from good to bad, or vice versa, or in that interleaving with other cycles may change). A cycle C is called *critical cycle for a reversal* in case that C has nodes both within and outside the reversed area. The following lemma summarizes why critical cycles deserve special attention.

Lemma 4.16. *Effects of reversals on being good or bad and on interleaving*
Every good cycle that is uncritical for a reversal remains good after the reversal. Every bad cycle that is uncritical for a reversal remains bad after the reversal. Every bad cycle that is critical for a reversal turns into a good cycle after the reversal. Note that a good cycle that is critical for a reversal may still be a good cycle after the reversal. For two cycles which are not both critical for a reversal, interleaving or non-interleaving is not affected by the reversal.

Proof. Easy exercise.

Corollary 4.17.
Execution of a reversal acting on two R-links of a cycle C of a bad component K does not lead to a loss or gain of cycles within K.

Now we can prove Lemma 4.15.

Proof. Consider some reversal acting on two R-links r and s of cycle C. We distinguish two cases. First, assume that the reversed area contains at least a further R-link t of C. The reversal changes orientation of t whereas orientations of r and s remain unchanged. Thus a good cycle results from C. Second, assume that the reversed area does not contain further R-links of C. Hence, r and s are the two adjacent R-links of a D-link d. By Lemma 4.23, d must be intersected by a further D-link d' belonging to a cycle C' in the same component as C. Thus C' is a bad cycle with some of its R-links within and some outside the reversed area. Thus, the reversal turns C' into a good cycle. Finally, we must show that the component of C is preserved. This follows from the following observations: interleaving of cycles is not changed by the reversal in case that we consider pairs of cycles that are not both critical for the reversal. Circles that are critical for the reversal interleave with C before and after the reversal since there is in any case a path between the border nodes of the reversed area. Thus, though interleaving between critical cycles may be changed by the reversal this does not affect membership of such cycles to K.   □

Lemma 4.15 leads to a first upper bound for the reversal distance d of a signed permutation, namely $n + 1 - k + b$ with n being the number of signed numbers, k the number of cycles, and b the number of bad components in the RD-diagram.

Lemma 4.18. *Simple upper bound*

$$d \leq n + 1 - k + b.$$

Interestingly, using as many extra reversals as there are bad components is not always necessary. To see under what conditions we can sort an RD-diagram with fewer extra reversals we need some further notion. What shall be defined is the *area of a component*. Areas of an RD-diagram are regions that are

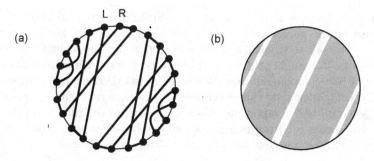

Fig. 4.92. (a) fine-structure of components; **(b)** shows components as "black box" areas

bounded by an alternating sequence of circle segments and lines. The example in Fig. 4.92 gives a first impression of what we mean by "areas".

The formal definition of the area of a component simply is as follows. Take the segments of the circle constituting the RD-diagram that exactly cover all nodes of the component. Traverse these segments counter-clockwise and always draw a line from the end node of each segment to the start node of the next segment. As a further example, consider an arrangement of four areas in Fig. 4.93. Assume that all of these four areas correspond to bad components. The indicated single extra reversal is sufficient to turn all bad components into good ones.

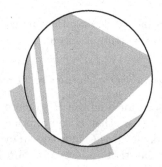

Fig. 4.93. Twisting all separating bad components at once into a good component

What happens in this example is that bad components which separate others may take advantage of applying any reversal acting on the separated components. Formally defined, we say that a component K *separates* two components K_1 and K_2 in case that K contains a D-link d such that the areas of K_1 and K_2 are on different sides of d. In the example above we had two components that separated others. Application of the indicated reversal melts the outer components and at the same time twists the separating components. Thus a common good component is generated by a single reversal. Of course, the number of cycles has decreased by 1. Spending a second extra reversal could restore the number of cycles. Taking everything together, we required only two reversals to transform four bad components into a single good component without changing the number of cycles. What obviously counts is the number of bad components that do not separate other two bad components. Such bad components are called *hurdles*. Clearly note that hurdles are defined to not separate bad components, whereas it is admitted that a hurdle separates a bad and a good component, or two good components. What has been observed for the example above can also be proved in general.

Lemma 4.19. *Component merging*
Consider two bad components K and K', cycle C in K and cycle C' in K'. Then, any reversal between an arbitrary R-link of C and an arbitrary R-link of C' turns K, K' and all components that separated K and K' into a single good component.

Proof. First, a reversal between an R-link of C and an R-link of C' melts cycles C and C' into a single good cycle. Since after the reversal the border nodes of the reversed area are joined by a path P, every component that separates K and K' belongs to the created common good component. Interleaving of cycles may change only for pairs of cycles that are critical for the reversal. Anyhow, such critical cycles interleave with P and thus belong to the new created component. Hence, the new created component contains nothing else than the components K and K' as well as all components that separated K and K'.   □

We next show that hurdles indeed deserve their name: every hurdle inevitably requires an extra reversal for getting it sorted.

Lemma 4.20.
Any reversal decreases the number of hurdles by at most 2.

Proof. There are two reasons for a hurdle H to loose the property of being a hurdle. First, it might be the case that H contains a cycle C that is critical for the reversal. Then the reversal turns C into a good cycle and H becomes part of a good component. We say that H looses its hurdle property by *twisting*. The diagrams in Fig. 4.94 show that at most two hurdles can be twisted by a reversal (given three bad components that are twisted by the reversal, one of them must separate the other two).

Second, it might be the case that hurdle H is not critical for the reversal, but after the reversal, component H separates two bad components. Since H did not separate bad components before the reversal, a new bad component must have been created by the reversal. Thus the location of H must be as in Fig. 4.95 (it must necessarily be located outside the reversed area; the reader may explain why H cannot be located within the reversed area).

We say that H looses its hurdle property by *separation*. As Fig. 4.95 shows there can be at most one hurdle that looses its hurdle property by separation.

Fig. 4.94. At most two hurdles are twisted into good components.

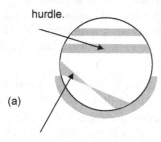
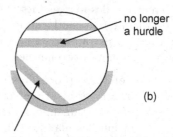

hurdle.

no longer
a hurdle

(a)

(b)

Good component that is critical.

A new bad component is created.

Fig. 4.95. Loosing a hurdle by separation: critical good component in (a) creates a new bad component in (b)

In addition, it cannot happen that hurdles are destroyed by twisting, and at the same time a further hurdle is lost by separation since in the former case there is no new bad component created. □

Lemma 4.20 allows us to state an improved lower bound for the reversal distance d of a RD-diagram of n signed numbers with k cycles and h hurdles.

Lemma 4.21. *Improved lower bound*

$$n + 1 - k + h \leq d.$$

Proof. We show that number $n + 1 - k + h$ decreases after a reversal by at most value 1. In case two hurdles H_1 and H_2 are destroyed by twisting, the executed reversal necessarily took place between R-links of different cycles. Thus, k decreases by 1. Hence $n + 1 - k + h$ decreases by 1. In case that exactly one hurdle H is lost by twisting, either the reversal took place between R-links of two different cycles or between R-links of a single cycle that must be a member of H, thus is a bad cycle. In the former case, k decreases by 1, hence $n + 1 - k + h$ is unchanged, in the latter case k is unchanged, hence $n + 1 - k + h$ decreases by 1. In case that exactly one hurdle is destroyed by separation it might be the case that the number of cycles decreases by 1. Fig. 4.95 shows that in case of separation at least one new hurdle is created. Thus the number of hurdles is not decreased at all and $n + 1 - k + h$ decreases at most by 1. □

We switch back to upper bounds and ask whether $n + 1 - k + h$ is sufficient to sort a RD-diagram of n signed numbers with k cycles and h hurdles. For this we recapitulate what procedures are available so far to transform hurdles into parts of good components.

- **Single hurdle destruction**: one extra reversal applied between two R-links of a cycle of a hurdle turns the hurdle into a good component and lets the number of cycles unchanged.
- **Hurdle merging**: one extra reversal applied to R-links of cycles in two different hurdles merges these hurdles as well as all components that separate them into a good component, but decreases the number of cycles by 1. A further extra reversal restores the number of cycles. Thus, two extra reversals removed two hurdles.

So far, both of these procedures sound good with respect to the required number of extra reversals. However, there is a final little obstacle in proving $n + 1 - k + h$ to be an upper bound for reversal distance d, namely that destruction of hurdles may lead to the creation of new hurdles. To see how this may happen look at the example in Fig. 4.96 with six bad components, three of them being hurdles (black shaded). Destroying either one of them or merging either two of them turns a non-hurdle (grey shaded) into a new hurdle.

Fig. 4.96. Whatever is done, a non-hurdle turns into a hurdle

This motivates the following definition. A hurdle is called *super-hurdle* if removing it from the RD-diagram turns a former non-hurdle into a hurdle. Fortunately, as long as there are at least four hurdles in a RD-diagram, we can easily avoid creation of new hurdles: always apply hurdle merging to non-neighbouring hurdles H_1 and H_2, i.e. hurdles having the property that between H_1 and H_2 as well as between H_2 and H_1 (counter-clockwise along the RD-diagram) there are further hurdles located (Fig. 4.97).

Whatever bad component H separated hurdles H_1 and H_2 before the reversal, H will either become part of the new generated good component or it will separate H_3 and H_4. Thus H either vanishes as a bad component or still remains a bad component that is not a hurdle. Things get critical if we are left with exactly three hurdles all of them being super-hurdles (Fig. 4.95). Whatever hurdles we destroy, a new hurdle is created. In such a situation that is

Fig. 4.97. Hurdle merging between non-neighbouring hurdles

called a *3-fortress* one additional extra reversal is required. RD-diagrams that finally lead to a 3-fortress after iterated hurdle merging can be characterized as follows: a RD-diagram is called a *fortress* if it contains an odd number of at least 3 hurdles, and all of its hurdles are super-hurdles. Indeed, under the presence of a fortress a further extra reversal is inevitably required to sort the diagram. For an RD-diagram define its *fortress number* f to be 1 if the diagram is a fortress, and 0 otherwise. Later it will be important to know that hurdle merging does not destroy fortresses, unless a 3-fortress is present.

Lemma 4.22. *Fortress preservation*
Hurdle merging of two non-neighbouring hurdles in a fortress with at least 5 hurdles preserves the fortress property.

Proof. This is obvious, as an odd number of hurdles is still present after application of such a reversal, and each of the non-merged super-hurdles is still a super-hurdle after the reversal. □

Lemma 4.23. *Best lower bound*

$$n + 1 - k + h + f \le d.$$

Proof. The lemma is proved by showing that any reversal decreases number $n+1-k+h+f$ by at most 1. In case $f = 0$ nothing has to be shown since we already know that number $n + 1 - k + h$ decreases by at most value 1. Now consider the case $f = 1$. Thus the RD-diagram has an odd number of at least 3 hurdles, and all of its hurdles are super-hurdles. If no hurdles are destroyed, we also know that the fortress is preserved, thus $n + 1 - k + h + f$ decreases by at most 1 (in case that k increases by 1). If a single hurdle is destroyed by twisting, a new hurdle is generated (by the super-hurdle property) and thus the presence of a fortress is preserved. Thus, $n + 1 - k + h + f$ will not decrease at all. The same applies if a single hurdle is destroyed by separation. If two hurdles are destroyed by twisting and we had at least five hurdles, the fortress property is preserved, thus again $n+1-k+h+f$ decreases by at most 1. If two hurdles are destroyed by hurdle merging in a 3-fortress we have no longer a fortress, thus f decreases by 1. Since for a 3-fortress hurdle merging creates a further hurdle (note that this is not the case for greater fortresses), $n + 1 - k + h + f$ again decreases by at most 1. □

Now we arrived at a lower bound that will turn out to be an upper bound for the minimum number of reversals required to sort a RD-diagram.

Lemma 4.24. *Best upper bound*

$$d \leq n + 1 - k + h + f.$$

Proof. Whenever there are good components these are sorted first. This does not affect hurdles and the fortress property. In case of a fortress, the number of hurdles is iteratively decreased by always applying hurdle merging between non-neighbouring hurdles. This preserves the fortress property until we arrive at a 3-fortress. Then we use the extra reversal provided by $f = 1$ and finish sorting by a hurdle merging with the available number of reversals. In case of a non-fortress we must take care not to generate a fortress (note that by $f = 0$ we have not available the extra reversal necessary for the destruction of a fortress). There are indeed RD-diagrams that are somehow critical. Look at the example shown in Fig. 4.98. There are five hurdles, two of them being super-hurdles. Thus the diagram is not a fortress. Though using the formerly safe strategy of merging two non-neighbouring hurdles, the reversal leads to a diagram with three hurdles all of which are super-hurdles, thus creates a fortress.

Fig. 4.98. Generation of a fortress

In case of an odd number of at least five hurdles, not all of which are super-hurdles, it is thus better to not merge two hurdles but instead to destroy (turn into a good component) a single hurdle that is not a super-hurdle. Turning a bad component into a good one is the same as removing it from the RD-diagram. As a non-super-hurdle is removed, there is no new hurdle created. The result is a diagram with an even number of hurdles, definitely not a fortress. Thus, $f = 0$ has been preserved.   □

Corollary 4.25. *Hannenhalli/Pevzner*

$$d = n + 1 - k + h + f.$$

All what is left is to finally show how to sort good components with the available number of reversals.

Exercise

4.2. The reader may treat all cases of 3 or less hurdles and show, for each case, how to sort with the available number of reversals.

4.5.3 Padding RD-Diagrams

As Fig. 4.83 shows, RD-diagrams may be rather complicated. Sorting good components is in addition complicated by the fact that not every reversal between R-links of a single good cycle of a good component G produces again good components from G. To show how a proper choice of the applied reversal is done producing again only good components, we must first simplify RD-diagrams considerably. The simplification is towards cycles with only 4 or 2 nodes by a process called *interleaved padding*. We describe this process first. Consider a cycle C having more than 4 nodes in a RD-diagram. We distinguish two cases.

Case 1

C has at least one pair of intersecting D-links d_1 and d_2. On cycle C we choose a D-link d different from d_1 and d_2 and an R-link r such that d_1 and d_2 separate d and r on cycle C (see Fig. 4.99). Then a pair of new nodes is introduced, R-link r is split into two R-links, and D-link d is distributed to the new nodes in such a way that two cycles of shorter length result, one with D-link d_1, the other one with D-link d_2. Note that a bad cycle C splits into two intersecting bad cycles, and a good cycle C splits two intersecting (due to D-links d_1 and d_2 being distributed to both new cycles) cycles with at least one of them being good. Thus, component and hurdle structure of an RD-diagram is left invariant by this process (the reader should convince himself at this point that further cycles interleaving with C either interleave with one of the new cycles, and interleaving between any other cycles than C does not change at all). What changes is number n of node pairs which increases by 1, number of cycles which also increases by 1, and length of cycles which gets shorter for the cycle C.

Case 2

There are no interleaving D-links on cycle C (Fig. 4.100). This means that C must be a bad cycle. Consider an arbitrary D-link d_1 on C. Applying the necessary condition on RD-diagrams that was proved above, d_1 must be intersected by some further D-link d', that thus belongs to another cycle C'. As C goes from on side of d' to the other (via D-link d_1), there must be a further D-link d_2 on C intersecting d'. As in the case before, choose some D-link d and R-link r on C which are located on different sides of line d'. As before, R-link r is split into two R-links using a pair of fresh nodes, and D-link d is distributed in a suitable manner to these new nodes such that two shorter

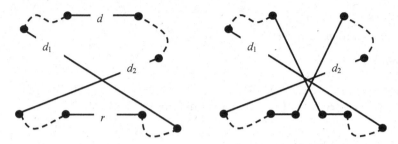

Fig. 4.99. Padding of a cycle with interleaving D-links

cycles result which both intersect D-link d'. The effect of the construction is that a common component results with bad cycle C being split into two shorter bad cycles. Thus quality of the component (being good or bad) is preserved. Again, component and hurdle structure of the RD-diagram is unchanged.

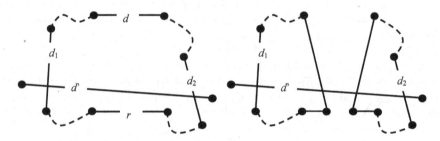

Fig. 4.100. Padding of a cycle without interleaving D-links

Lemma 4.26.
Number $n+1-k+h+f$ is not affected by an interleaved padding. For the RD-diagram obtained after application of an interleaved padding, reversal distance has not got smaller.

Proof. As n grows by the same value as k, $n-k$ is unchanged. Parameters h and f are not altered since no new components are created and quality of existing components is not altered. To show that original reversal distance d is not greater then reversal distance d_{padd} after padding, it is useful to reinterpret RD-diagrams as representation of signed permutations. In the example diagram in Fig. 4.101 we consider the RD-diagram of signed permutation $+2$ -5 -3 -1 $+4$. The diagram resulting after the executed padding can be interpreted as RD-diagram of the extended signed permutation $+2$ -5 $+1\frac{1}{2}$ -3 -1 $+4$. Here we made use of signed rational numbers which does not cause any problems. The reader being fixed on the usage of signed integers may renumber rational numbers and work with signed permutation $+3$ -6 $+2$ -4 -1 $+5$.

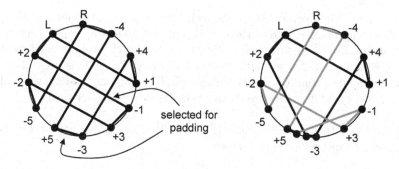

Fig. 4.101. Padding is insertion of new signed numbers

Consider a sorting of the padded sequence of signed numbers using d_{padd} reversals. Deleting any occurrence of $+1\frac{1}{2}$ and $-1\frac{1}{2}$, we get a sorting of the original sequence of signed numbers using at most d_{padd} reversals ("at most" since there may be reversals working only on $+1\frac{1}{2}$ and $-1\frac{1}{2}$; after deletion of $+1\frac{1}{2}$ and $-1\frac{1}{2}$ there result "dummy" steps without any effect). As d is the least required number of reversals, we get $d \leq d_{\text{padd}}$.   □

Note that we finally must prove the *Main Lemma* only for padded diagrams with cycles of length 2 or 4. After having shown $d_{\text{padd}} \leq n + 1 - k + h + f$, we infer $d \leq n + 1 - k + h + f$ using the estimation $d \leq d_{\text{padd}}$ shown before. The converse estimation $n + 1 - k + h + f \leq d$ has been shown already in the sections before, even for arbitrary (also non-padded) diagrams.

4.5.4 Sorting Padded Good Components

From now on we assume that we deal with RD-diagrams consisting of cycles of length 2 or 4 only. To simplify presentation of diagrams trivial cycles of length 2 are never drawn.

Lemma 4.27. *Main Lemma*
Every good component K contains at least one good cycle such that the (unique) reversal between its two R-links splits it into two trivial cycles, as we already know, and does not generate any bad components from K.

Proof. Choose an arbitrary good cycle C within K. The reversal between its two R-links is briefly called *C-reversal*. For any cycle X that is different from C we denote by X^C the cycle resulting from X by the C-reversal. Likewise, for any set of cycles K not containing C we denote by K^C the set of cycles X^C with X in K. A cycle interleaving with C is briefly called a *C-critical cycle*. The following is easily verified. C-critical cycles, and only these, change their character (being good or bad) after application of the C-reversal. For pairs of C-critical cycles interleaving or non-interleaving changes after application of the C-reversal, whereas for pairs of cycles with at most one C-critical interleaving or non-interleaving is not altered. Executing the C-reversal splits cycle

C into two trivial cycles. Now assume that the C-reversal generated at least one bad component K^C from the component of C. Thus K^C must contain at least one cycle D^C such that D is C-critical. Since K^C was assumed to be a bad component, we know that D^C is a bad cycle. Thus, D is a good cycle. By the necessary condition on RD-diagrams, bad cycle D^C interleaves with a further bad cycle E^C. Now we distinguish whether E is C-critical.

Case (1) is that E is C-critical (Fig. 4.102). Thus E is a good cycle. Furthermore, D and E do not interleave.

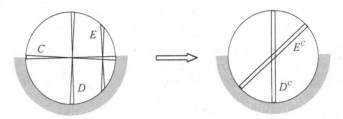

Fig. 4.102. E is C-critical

Case (2) is that E is not C-critical (Fig. 4.103). Thus E is a bad cycle. Furthermore, D and E interleave.

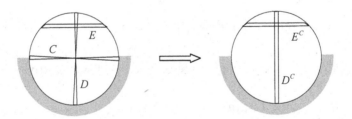

Fig. 4.103. E is not C-critical

Instead of C-reversal we execute D-reversal and show that the number of cycles in all bad components it produces is less than the number of cycles in all bad components that are produced by the C-reversal. As above, we use the notion X^D to denote the cycle resulting from cycle X by the D-reversal (for a cycle X that is different from D). The result is shown for both cases in Fig. 4.104, (1) for the first case, and (2) for the second case.

In both cases we observe that after D-reversal cycles C^D and E^D are part of a good component whereas after C-reversal cycles D^C and E^C were part of a bad component. Thus, concerning cycles C, D, E, D-reversal outper-

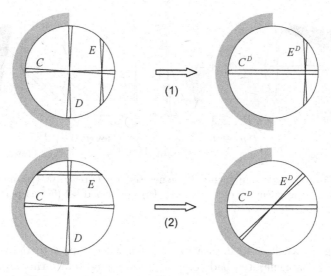

Fig. 4.104. Effect of D-reversal instead of C-reversal

forms C-reversal with respect to the number of cycles in freshly generated bad components.

But we must also consider bad components that are eventually freshly generated by D-reversal and show that these also are part of bad components after C-reversal. We show this in Fig. 4.105 for the first case (second case is to be treated similarly and left to the reader). A bad component B^D that is freshly generated by D-reversal does not interleave with E^D or C^D, since the latter cycles are part of a good component. Thus, it is located within one of the 4 sectors (upper left, upper right, lower right, lower left) into which the RD-diagram is partitioned by cycles E^D or C^D. We concentrate on the upper left sector and leave it to the reader to draw corresponding diagrams for the other three sectors. Now also consider cycle D and B as it was before D-reversal. A simple case is that B does not interleave cycle D. In that case neither D-reversal nor C-reversal affects component B. Thus, B^C contributes as many bad cycles to the diagram after C-reversal as B^D contributed to the diagram after D-reversal. A more critical case is when B interleaves cycle D. Then all cycles in B^D that interleave D correspond to good cycles in B that interleave cycle C. Thus, B may consist of a mixture of good and bad cycles. After C-reversal all good cycles of B are twisted back again into bad cycles of B^C. Also, for every pair of cycles in B^D for which interleaving was altered in B, interleaving is altered again in B^C. This means that B^C again is a connected set of bad cycles that is part of a bad component freshly generated by C-reversal. Summarizing, whatever cycles in fresh bad components are generated by C-reversal, they are also generated in bad components by D-reversal. Thus, C-reversal indeed performs strictly better than D-reversal with respect to freshly generated bad cycles.

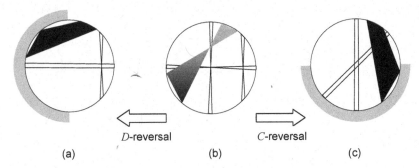

Fig. 4.105. (a) Bad component B^D generated by D-reversal; (b) mixed set B of cycles before application of D-reversal; (c) connected set B^C as part of a bad component after C-reversal

Having shown how to improve a reversal with respect to the number of cycles in freshly generated bad components we may iterate this process until we finally arrive at a reversal between R-links of a good cycle which does not create any fresh bad components at all.   □

4.5.5 Summary of the Sorting Procedure

Before sorting starts, we transform the considered RD-diagram into a simpli-fied one containing only cycles of length 2 and 4. There is one step described in the sections before that must be slightly modified in order to preserve lengths 2 and 4 of cycles in the diagram. It is the step of hurdle merging. Here, two bad cycles of length 4 are merged into a single good cycle of length 8. Imme-diately applying two interleaved paddings produces three interleaving good cycles of length 4 (Fig. 4.106). The complete procedure runs as follows.

> Pad the RD-diagram into one with cycles of length 2 or 4;
> **while** there are non-trivial cycles **do**
>   **if** there are good components **then**
>   sort these according to *Main Lemma*
>   **else if** there is an even number of at least 4 hurdles **then**
>   apply hurdle merging on non-neighbouring hurdles
>   **else if** there is a fortress with at least 5 hurdles **then**
>   apply hurdle merging on non-neighbouring hurdles
>   **else if** there is an odd number of at least 5 hurdles **then**
>   destroy a single non-super-hurdle
>   **else**
>   sort the diagram with at least 3 hurdles
>   **end if**
> **end while**

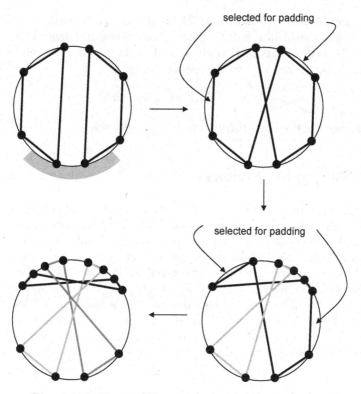

Fig. 4.106. Two paddings applied after hurdle merging

Choosing in every step a reversal that decreases the value of $n + 1 - k + h + f$ requires some efforts: paddings must be used; sorting good components is rather involved. As a consequence, an estimation of the time complexity of this algorithm is not easy. We may alternatively proceed as follows. As we know that a given RD-diagram (not necessarily padded) with k cycles, h hurdles and fortress number f can be sorted in $n + 1 - h + k + f$ reversals, we know that in every step there always exists a reversal that decreases this value by 1. We simply may try out each possible reversal and calculate for each such reversal the updated value of $n + 1 - k + h + f$, until one is found that decreases this value by 1. Since at most $2n$ reversals are sufficient to sort a signed permutation of n numbers, at every step there at most n^2 reversals to be tried out, and computation of term $n + 1 - k + h + f$ requires $O(n^2)$ time - the reader may work out details of the computation of cycles, being good or bad, interleaving of cycles, components, hurdles, super-hurdles, fortresses, and estimate time required to compute all these things - we arrive at an $O(n^5)$ algorithm.

We conclude this section with a look at the example of a sorting in 26 steps presented in Chap. 2 in Fig. 2.15. The corresponding RD-diagram developed at the beginning of this section showed 11 cycles, no hurdles, and thus no fortress. Thus, the reversal distance evaluates to

$$36 + 1 - 11 + 0 + 0 = 26$$

showing that the executed sorting indeed was optimal.

4.6 Bibliographic Remarks

PQ-trees were introduced in Booth & Lueker [13]. An extended presentation of various algorithms for the construction of suffix trees, and also a huge collection of applications of suffix trees, in particular combined with the least common ancestor algorithms described in Sect. 4.4 can be found in Gusfield [31]. Genome rearrangement is covered by Pevzner [64] and Setubal & Meidanis [68]. The idea of using reality-desire diagrams to visualize genome rearrangement problems was proposed in Setubal & Meidanis [68] as an alternative to the less easy to grasp diagrammatic presentations in Pevzner [64], though the former book leaves out proofs of the main combinatorial parts of the correctness proof of the Hannenhalli/Pevzner formula.

5

NP-Hardness of Core Bioinformatics Problems

The reader familiar with basic notions of complexity theory, in particular with the concept of NP-completeness, may skip over Sect. 5.1 (or consult this section whenever necessary). By the same way, being well experienced in reduction techniques the reader may also skip over Sects. 5.2 and 5.3. To obtain a self-contained presentation of NP-complete bioinformatics problems we have indeed shown NP-completeness of all intermediate problem that are used in reductions of this chapter, starting from Cook's famous 3SAT problem.

5.1 Getting Familiar with Basic Notions of Complexity Theory

What is an "algorithmic problem" and what does it mean to efficiently solve an algorithmic problem?

Most algorithmic problems occurring in bioinformatics are optimization problems. Such consist of the following ingredients:

- *Alphabet* of characters Σ
- *Instance subset I*: a polynomially decidable subset of Σ^*
- *Solution relation S*: a polynomially decidable and polynomially balanced relation[1] over $\Sigma^* \times \Sigma^*$; for simplicity we assume within this chapter that for every instance $x \in I$ there exists at least one y with $(x, y) \in S$.
- *Cost function c* that assign to each pair $(x, y) \in S$ an integer (or rational) value $c(x, y)$
- *Mode* min or max

To *solve* such a minimization (maximization) problem means to develop an algorithm A that computes for every instance x a solution $y = A(x)$ for x with

---

[1] This means that for all $(x, y) \in S$, the length of y is bounded by a polynomial in the length of x.

minimum (maximum) costs $c(x, y)$. To efficiently solve a problem requires in addition that the computing time on input x is bounded by some polynomial in the length of input instance x.

The meaning of the components of an optimization problem is as follows: usually, among all strings over alphabet Σ there are lots of strings that do not encode meaningful or admissible input instances. For example, $\wedge \wedge \neg x$ does not encode a Boolean formula, 0001402 usually is not allowed as representing an integer due to leading zeroes, 0 is not admissible in case of integer division, -325 is not admissible for computing square roots. Instance set I expresses which strings over Σ represent meaningful or interesting input instances of a problem. Of course it should be efficiently testable whether a given string fulfils the requirements of instance set I. Given instance x, we are interested in computing so-called *optimal solutions*, that is, strings y with $(x, y) \in S$ and minimal or maximal value $c(x, y)$. In order for solutions to be efficiently constructible, their lengths should necessarily be bounded by a polynomial in the length of input x (otherwise already writing down a solution would require exponential space and time, not to speak about the complexity of finding it). For every proposed string y it should be efficiently testable whether it is a solution for x. Thus, the "only" problem with an optimization problem is that although simple browsing through the space of all solution for an instance x is possible, this leads to exponential running time. The main theme of complexity theory is to find out for a concrete optimization problem whether there is a more intelligent, thus efficient, algorithm, or whether exponential execution time is inherent to the problem and thus (with high confidence) unavoidable.

How did we show so far that an algorithmic problem is efficiently solvable?

In theory, this is easy: simply develop an algorithm that solves the problem and runs in time polynomial in input length. Practically, it may be rather intricate to find an efficient algorithm, provided such an algorithm exists at all. One goal of this book is to present and illustrate with selected examples a few general principles (paradigms) that often guide us towards efficient algorithms.

How can we show that an algorithmic problem is not efficiently solvable?

This requires deeper theoretical preparations. First, optimization problems are transformed into languages which are used in complexity theory to encode problems. This simplification proceeds in several steps. First, an optimal solution y for a given problem instance x usually can be obtained on basis of an algorithm which only computes the optimal value of the cost function that can be achieved with any solution for problem instance x. The strategy is to successively fix the bits of solution y by suitably modifying parts of instance x and asking whether the former optimal value is unaltered; depending on the obtained answer we usually may infer knowledge about one or more bits of y.

As an example, consider TSP, the travelling salesman problem. Let c be the minimum possible cost of a Hamiltonian cycle. Now start with an arbitrary node v. For a proposed successor node w of v, set the cost of edge (v, w) to infinity (thus suppressing this edge) and compute the optimal value c' after this suppression. If $c' = c$ holds we know that edge (v, w) is not required for an optimal Hamiltonian cycle, so we may suppress it forever and try out the next successor node of v. If $c' > c$ holds we know that edge (v, w) necessarily must be part in any possible optimal Hamiltonian cycle. So we restore cost of edge (v, w) to its original value and proceed to determine the successor of node w in an optimal Hamiltonian cycle by the same way as the successor of node v was determined. Thus, if the given graph has n nodes, at most n^2 calls of an algorithm computing optimal values leads us to a cheapest Hamiltonian cycle.

Next, computing optimal values could be done if we had an algorithm that gives, for an instance x and integer bound b, the answer to the *question* whether a solution y exists with cost $c(x, y)$ at most b in case of mode = min, or a solution exists with cost $c(x, y)$ at least b in case of mode = max. Having such a decision algorithm and observing that the lengths of solutions are polynomially bounded in the length of the input, binary search allows efficient navigation towards the optimal value of the cost function.

Finally, questions whether a solution y exists with costs at most or at least b could be answered if we had an algorithm which decides whether pair (x, b) consisting of instance x and bound b belong to the following language:

$$L = \{(x, b) \mid x \in I \wedge \exists y \text{ with } (x, y) \in S \wedge c(x, y) \le b \text{ resp. } c(x, y) \ge b\} .$$

Here, we only have to note that, given input string x, a preprocessing test whether x belongs to I, must be incorporated into an algorithm.

So we end with a language L that encodes the original optimization problem. Showing that the original optimization problem does not admit an efficient (polynomial time) algorithm is thus reduced to showing that the encoding language L does not admit an efficient algorithm. For this latter task the notion of polynomial reductions between languages and the notion of NP-completeness is established.

How are complexity class NP and NP-completeness defined, and what is the meaning of NP-completeness?

Complexity class NP consists of all languages L that are definable in existentially quantified form

$$L = \{x \mid \exists \text{ some } y \text{ such that } (x, y) \in S\}$$

with a polynomially decidable and polynomially balanced relation S. Note that the language used above to represent an optimization problem is a language in NP. Informally stated, complexity class NP reflects the well-known generate-test problem-solving paradigm with a search space of exponential size and a polynomial time test. Almost all practically relevant problems with the exception of a few undecidable problems occurring in mathematics belong to class NP.

Languages L_1 and L_2 may be compared with respect to their inherent complexity by *polynomially reducing* L_1 to L_2, formally noted $L_1 \leq_{\text{pol}} L_2$, meaning that a polynomial time computable function f exists such that for all strings x, $x \in L_1$ is equivalent to $f(x) \in L_2$. Thus, reducing L_1 to L_2 means that we may transform questions $x \in L_1$ into equivalent questions $f(x) \in L_2$, or that L_1 is in some sense contained within L_2 as shown in Fig. 5.1.

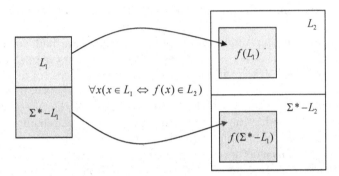

Fig. 5.1. The meaning of reductions

A more general sort of reduction, called *truth-table reduction*, allows a single question $x \in L_1$ to be reduced to a Boolean combination of several questions $f_1(x) \in L_2, \ldots, f_m(x) \in L_2$. The following statement holds for both reduction concepts and follows immediately from the definitions.

Lemma 5.1.
If $L_1 \leq_{\text{pol}} L_2$ and L_2 is polynomially solvable, then L_1 is polynomially solvable, too. In this sense, L_1 is at most as difficult to solve as L_2 is. Stated the other way around, if L_1 is not polynomially solvable, then L_2 is not polynomially solvable, either. If $L_1 \leq_{\text{pol}} L_2$ and $L_2 \leq_{\text{pol}} L_3$, then $L_1 \leq_{\text{pol}} L_3$.

Proof. Let L_1 be reduced to L_2 via reduction function f. Having a polynomial algorithm for L_2, we must simply preprocess input x to L_1 by computing $f(x)$ and then invoke a polynomial algorithm for L_2 on $f(x)$ to obtain a polynomial algorithm for L_1. It is to be used here that the length of $f(x)$ is bounded by a polynomial in the length of x whenever computation time of e.g. a Turing machine (see next section for a description of Turing machines) computing $f(x)$ is bounded by a polynomial in the length of x. Transitivity of relation $\leq_{\text{pol}}$ requires two reductions to be executed consecutively.   □

Now a language L is called *NP-complete* if it is an NP-language and every NP-language L' may be polynomially reduced to L. Thus, NP-complete languages are greatest elements of NP with respect to the partial ordering $\leq_{pol}$, thus they are the most difficult languages in NP.

Theorem 5.2.
Let L be an NP-complete language. If L is polynomially solvable then P = NP follows.

Proof. Every NP-language L' is polynomially reducible to NP-complete language L. Polynomial solvability of L thus would give polynomial solvability of every NP-problem, thus P = NP. □

Remark 5.3.
At least 50 years (or more) of work on thousands of NP-complete problems by ten thousands (or more) researchers have not resulted in a polynomial algorithm for any of these problems. Many people take this as strong evidence that P $\neq$ NP, though there are also a few people expecting P = NP, and some expecting the P = NP question to be a question that, like the axiom of choice in axiomatic set theory, cannot be decided on basis of usual foundations of computer science (Peano's axioms, or axioms of set theory). Showing that a language is NP-complete is thus, for most people, strong evidence that L cannot be polynomially solved.

What can be done in case a problem is NP-complete?

Most often we will develop approximation algorithms in case of optimization problems whose underlying language is shown to be NP-complete. Such return solutions that are at most a predefined constant factor worse than an optimal solution. A different relaxation would be to use probabilistic algorithms which return a correct solution with some predefined confidence. The examples presented in this book do not make use of probabilistic algorithms. A further option is to use adaptive algorithms, e.g. neural networks or hidden Markov models, both being able to solve a problem by being trained to training instances. We will present a couple of problems that are "solved" this way. Adaptive algorithms are particularly suitable in domains with some lack of theoretical insights or by the presence of error within data, as it is often the case in bioinformatics problems. Finally, various heuristics play a role in bioinformatics problem solving.

5.2 Cook's Theorem: Primer NP-Complete Problem

Are there any NP-complete languages at all? Cook and Karp (1971) showed that the satisfiability problem SAT for Boolean logic is NP-complete [22, 38]. SAT is the language of all Boolean formulas which can be made 'true' by a suitable assignment of truth values to its Boolean variables. Being in NP is obvious for SAT. To establish NP-completeness forces us to get a little bit more formal with respect to the notion of an algorithm. It is most recommendable to use Turing machines as abstract machine for the implementation of algorithms. Turing machines are state controlled machines that work on a tape containing cells numbered ... -2, -1, 0, +1, +2, ... Cells may contain characters from the alphabet of the Turing machine. At every moment of work, a Turing machine is in a certain inner state q and its head is placed at a certain cell containing some character a (see Fig. 5.2), empty cells being indicated by using the dummy symbol $\emptyset$. As long as the inner state is different from one of the final states 'yes' or 'no', a well-defined action takes place which is controlled by the transition function δ of the Turing machine. Given $\delta(q,a) = (a',m',q')$ with character a', move m' (which may be 'L', 'S' or 'R'), and next state q', the Turing machine prints character a' onto the actually scanned cell, moves left to the next cell in case $m' = $ 'L', remains at the actual cell in case $m' = $ 'S', or moves right to the next cell in case $m' = $ 'R'. Input string x of length n is initially written into cells 1 to n, control starts in initial state s, cells other than $1, \ldots, n$ are empty in the beginning.

Now let an arbitrary NP-problem L over alphabet Σ be given. Fix an enumeration of characters $a(1), \ldots, a(k)$ of alphabet Σ. Let L be given as

$$L = \{x \mid \exists y \text{ such that } (x,y) \in S\} \ .$$

Assume that Turing machine M as described above decides relation S in polynomial time $p(n)$, with n being the length of input (x,y). We may modify Turing machine M in such a way that after reaching one of the final states, 'yes' or 'no', M does not halt but remains in the reached final state without altering the tape. This can be achieved by introducing instructions $\delta(\text{yes},a) = (a,S,\text{yes})$ and $\delta(\text{no},a) = (a,S,\text{no})$. Assume that S is polynomially balanced, that is, for each x of length n and y with $(x,y) \in S$ the length of y is bounded by polynomial $q(n)$. We may even demand that for every $(x,y) \in S$ with x having length n, the length of y exactly equals $q(n)$. To achieve this, replace in the representation of language L solution relation S by

$$S' = \{ (x,z) \mid \text{length of } z \text{ is exactly } q(n) \text{ with } n \text{ being the length of } x$$
$$\text{and for some prefix } y \text{ of } z \text{ we have } (x,y) \in S \} \ .$$

Using S' instead of S defines the same language L and furthermore fulfils the desired stronger length requirement.

Now we may describe the polynomial reduction of L to SAT. For this, let be given string x of length n, say $x = x(1)x(2)\ldots x(n)$. By definition, $x \in L$ if and only if there is a string y of length $q(n)$ such that Turing machine M starting in state s on tape description $x\emptyset y$ arrives after exactly $T(n) = p(n+1+q(n))$ steps in state 'yes' and thus visits at most cells in the range of $-T(n)$ and $+T(n)$. This can be expressed as satisfiability of a certain Boolean formula φ_x depending on x which we construct next. For this, we simply describe in logical terms all what happens on tape cells $-T(n) \ldots +T(n)$ within $T(n)$ time steps. Initial situation is as shown in Fig. 5.3. We use the following Boolean

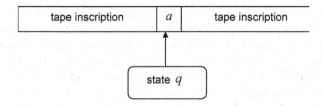

Fig. 5.2. "Architecture" of Turing machines

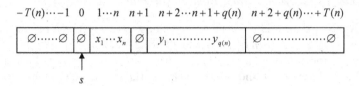

Fig. 5.3. Initial Turing machine configuration

variables that are in a suggestive manner denoted as follows.

$S_{t,q}$ *"after step t machine M is in state q"*
$P_{t,i}$ *"after step t working position of machine M is at cell i"*
$C_{t,i,a}$ *"after step t cell i contains character a"*

Initial situation is described by:

$S_{0,s}$
$P_{0,0}$
$C_{0,i,\emptyset}$ for $i \in \{-T(n), \ldots, -1, 0, n+1, n+2+q(n), \ldots, T(n)\}$
$C_{0,i,x(i)}$ for $i \in \{1, \ldots, n\}$
$(C_{0,i,a(1)} \vee C_{0,i,a(2)} \vee \ldots \vee C_{0,i,a(k)})$ for $i \in \{n+2, \ldots, n+1+q(n)\}$

The latter disjunctive formulas express that cells $n+2, \ldots, n+1+q(n)$ contain arbitrary characters; this is the equivalent to the phrase "there exists a string y of length $q(n)$" in the representation of language L. Proceeding of the computation from step $t < T(n)$ to step $t+1$ is described by taking for every entry $\delta(q,a) = (a', m', q')$ of the transition function of M the following logical implications, for all i and j between $-T(n)$ and $+T(n)$ with $i \neq j$ and all characters b from alphabet Σ:

$(S_{t,q} \wedge P_{t,i} \wedge C_{t,i,a}) \rightarrow S_{t+1,q'}$ next state is q'
$(S_{t,q} \wedge P_{t,i} \wedge C_{t,i,a}) \rightarrow C_{t+1,i,a'}$ character a' is written on cell i
$(P_{t,i} \wedge C_{t,j,b}) \rightarrow C_{t+1,j,b}$ no further changes on tape
$(S_{t,q} \wedge P_{t,i} \wedge C_{t,i,a}) \rightarrow P_{t+1,i-1}$ if $m' = L$ and thus $-T(n) < i$
$(S_{t,q} \wedge P_{t,i} \wedge C_{t,i,a}) \rightarrow P_{t+1,i}$ if $m' = S$
$(S_{t,q} \wedge P_{t,i} \wedge C_{t,i,a}) \rightarrow P_{t+1,i+1}$ if $m' = R$ and thus $i < T(n)$

Halting in final state and uniqueness of final state are described by literals $S_{T(n),\text{yes}}$ and $\neg S_{T(n),\text{no}}$.

The reader who wonders why we did not demand uniqueness of states, working position and content of every cell at every time step, may think of this question while reading the proofs below. The desired formula φ_x is the conjunction of all formulas constructed above. Note that φ_x consists of $O(T(n)^3)$ many sub-formulas, $T(n)^3$ resulting from the combination of any time t with any combination of working positions i and j.

Lemma 5.4.
If x belongs to language L then Boolean formula φ_x is satisfiable.

Proof. Assume that x is element of L. Choose some string y of length $q(n)$, say $y = y(1)y(2)\ldots y(q(n))$, such that Turing machine M when started in state s on tape inscription $x\emptyset y$ ends with state 'yes' after $T(n)$ steps. For every time t between 0 and $T(n)$, every working position i between $-T(n)$ and $+T(n)$, every Turing machine state q, and every character $a \in \Sigma$ assign truth values as follows: Boolean variable $S_{t,q}$ is assigned 'true' in case that M, when started in state s on tape inscription $x\emptyset y$, is in state q after time t, otherwise $S_{t,q}$ is assigned truth value 'false'. Boolean variable $P_{t,i}$ is assigned 'true' in case that M, when started in state s on tape inscription $x\emptyset y$, has working position i after time t, otherwise $P_{t,i}$ is assigned truth value 'false'. Boolean variable $C_{t,i,a}$ is assigned 'true' in case that, starting in state s on tape inscription $x\emptyset y$, cell i carries character a after time t, otherwise $C_{t,i,a}$ is assigned truth value 'false'. Call this the *truth value assignment describing M's computation on $x\emptyset y$* and denote it by $\Im_{x,y}$. It is obvious that $\Im_{x,y}$ makes all formulas constructed above 'true', since these formulas correctly describe what is going on during computation of M on initial tape inscription $x\emptyset y$. Thus, formula φ_x is satisfiable. □

Lemma 5.5.
If formula φ_x is satisfiable then string x belongs to language L.

Proof. Take a truth-value assignment $\Im$ that satisfies all formulas constructed above. Interpreting the disjunctions used in the first group, fix for every i between $n+2$ and $n+1+q(n)$ some character $y(i)$ such that $\Im$ assigns truth-value 'true' to $y(i)$. This defines a string y of length $q(n)$. Now consider the computation of length M that Turing machine M executes when started in state s on tape inscription $x\emptyset y$ and the truth value assignment $\Im_{x,y}$ constructed in the proof of the former lemma. For every time t the following inclusion relation between $\Im$ and $\Im_{x,y}$ is easily shown: every Boolean variable that is satisfied by $\Im_{x,y}$ is satisfied by $\Im$, too. This is clear for $t = 0$. Having shown this assertion for some t, it immediately follows for $t + 1$ since the formulas in the second group correctly describe what happens with state, working position, and tape content when going from time t to time $t + 1$. Note that so far we cannot exclude that $\Im$ satisfies more Boolean variables than $\Im_{x,y}$ does since there were no constraints other than the one on the uniqueness of final state arrived at time $T(n)$. This does not matter since the latter is all we require.

We know that $\Im$ satisfies Boolean variable $S_{T(n),\text{yes}}$, thus, by the uniqueness constraint, does not satisfy Boolean variable $S_{T(n),\text{no}}$, thus, by the inclusion relation shown above, $\Im_{x,y}$ does not satisfy Boolean variable $S_{T(n),\text{no}}$, thus Turing machine M started on $x\emptyset y$ does not end with state 'no', thus Turing machine M started on $x\emptyset y$ ends with state 'yes', thus x belongs to language L. □

5.3 "Zoo" of NP-Complete Problems

How is NP-completeness established for further NP-problems? This is simply done by reducing a known NP-complete language L to the actual problem L' under consideration. Immediately from the definitions it then follows that L' is NP-complete, too. Concrete reductions $L_1 \leq_{\text{pol}} L_2$ may range from rather trivial to very sophisticated depending on whether languages L_1 and L_2 are conceptually similar or complete dissimilar. A reduction $L_1 \leq_{\text{pol}} L_2$ is usually simple in case that L_2 provides a lot of structure that can be used to embed the concepts of L_1. Language SAT that was shown to be NP-complete in Sect. 5.2 is an example of a problem from a highly structured domain, Boolean logic.

> **SAT**
> Given Boolean formula φ, decide whether it is satisfiable, that is, whether there is an assignment of truth values to its variables that makes φ 'true'.

As Boolean logic can be seen as a formalization of a fragment of natural language, it is not astonishing that any natural language problem specification may be easily expressed in Boolean logic, as it was done in the proof of Cook's theorem. This, on the other side, tells us that using SAT to derive further NP-completeness results requires more efforts. Having a problem L in a rather weakly structured environment, any attempt to reduce SAT to L is expected to be more complicated. Nevertheless, in the beginning, nothing else can be done than reducing SAT to further NP-problems. But on the long run we will have a zoo of various NP-complete problems such that, given a further NP-problem L that is to be proved to be NP-complete, there is a good chance to find a similar problem L' within that zoo with a simple reduction from L' to L.

Starting with SAT, lots of further NP-completeness results are derived next. 3SAT is the following sublanguage of SAT consisting of all satisfiable Boolean formulas of the form

$$(L_{11} \vee L_{12} \vee L_{13}) \wedge (L_{21} \vee L_{22} \vee L_{23}) \wedge \ldots \wedge (L_{n1} \vee L_{n2} \vee L_{n3})$$

called formulas in *3-conjunctive normal form*, or formulas in *3-clausal form*, with so-called *literals* L_{ij}, that is Boolean variables (*positive literals*) or negated Boolean variables (*negative literals*).

> **3SAT**
> Given Boolean formula φ in 3-clausel form, decide
> whether it is satisfiable.

Note that the structure of formulas in conjunctive normal form is a conjunction of disjunctions of three literals each. Each disjunction is also called a *3-clause*. Also note that literals in a 3-clause need not be different. Admitting, at the moment, more than 3 literals within clauses, it is easily shown (using standard transformation rules for Boolean logic known as de Morgan's rules) that every Boolean formula can be transformed into an equivalent Boolean formula in conjunctive normal form. By the same way, every Boolean formula can be transformed into an equivalent Boolean formula in so-called *disjunctive normal form*, which for the case of 3 literals per disjunction are of the form

$$(L_{11} \wedge L_{12} \wedge L_{13}) \vee (L_{21} \wedge L_{22} \wedge L_{23}) \vee \ldots \vee (L_{n1} \wedge L_{n2} \wedge L_{n3}) \ .$$

From a complexity theory point of view, disjunctive and conjunctive normal forms are radically different: whereas we will soon show that satisfiability for formulas in conjunctive normal form is NP-complete, it is easily shown that satisfiability for disjunctive normal forms is polynomial solvable (in quadratic time or even time $O(n \log n)$ if we make use of fast sorting algorithms; the reader may think about details). This is in no contradiction to the before stated possibility to transform any formula either in conjunctive or disjunctive normal form; the reason is that transforming via de Morgan's rules conjunctive to disjunctive normal form (or vice versa) requires exponential time (even expands formula to exponential size).

Theorem 5.6.
3SAT is NP-complete.

Proof. Looking at the formulas we have used in Sect. 5.2 in the proof of NP-completeness of SAT we see that these are either positive literals, or disjunctions, or implications of the form $(L_1 \wedge \ldots \wedge L_n) \rightarrow L$ with literals $L_1, \ldots, L_n, L$, the latter formula being equivalent to the disjunction $(\neg L_1 \vee \ldots \vee \neg L_n \vee L)$. What remains to be done is to replace disjunctions with less or more than 3 literals by 3-disjunctions without affecting satisfiability. Disjunctions with less than 3 literals are simple expanded to disjunctions with 3 literals by simply repeating literals one or two times. A disjunction with more than 3 literals, say $(L_1 \vee L_2 \vee L_3 \vee L_4 \vee \ldots \vee L_n)$ is processed as follows: introduce a new Boolean variable L that serves as an abbreviation for $(L_1 \vee L_2)$ thus reducing the formula above to $(L \vee L_3 \vee L_4 \vee \ldots \vee L_n)$ containing one literal fewer than the formula before. Of course, it must be logically encoded that L stands for $(L_1 \vee L_2)$, meaning that we have to incorporate the implications $(L_1 \vee L_2) \rightarrow L$ and $L \rightarrow (L_1 \vee L_2)$. Transforming implications into disjunctions leads to $\neg(L_1 \vee L_2) \vee L$ and $(\neg L \vee L_1 \vee L_2)$, thus to $(\neg L_1 \wedge \neg L_2) \vee L$ and $(\neg L \vee L_1 \vee L_2)$, thus finally to 3 extra clauses $(\neg L_1 \vee L)$, $(\neg L_2 \vee L)$, and

$(\neg L \vee L_1 \vee L_2)$. The latter disjunction also makes clear why it is not possible to reduce the number of literals in clauses below 3. $\qquad\square$

Our next NP-complete languages come from the field of graph theory. Graphs (V, E) are either *undirected graphs* with node set V and set E consisting of undirected edges (links) between certain pairs of nodes, or *directed graphs* with directed edges (arrows) between certain pairs of nodes. Edges of an undirected graph (V, E) with node set $V = \{v_1, \ldots, v_n\}$ are usually encoded in form of an adjacency matrix which contains for every row i and column j bit 1 in case that there is an edge between nodes v_i and v_j, and bit 0 otherwise (the definition for directed graphs is similar). Alternatively, edges may be encoded for undirected graphs by listing, for every node v all nodes w with an edge between v and w (so-called *adjacency list*). A *clique* C in an undirected graph (V, E) is a subset of V such that any two different nodes in C are connected by an edge. Problem CLIQUE is the problem of finding a clique of maximum size. Formulated as a decision problem, CLIQUE is the language of all undirected graphs (V, E) together with an integer bound b such that (V, E) has a clique of size at least b.

CLIQUE
Given an undirected graph (V, E) and lower bound b, is there a clique of size at least b.

Theorem 5.7.
CLIQUE is NP-complete.

Proof. 3SAT is polynomially reduced to CLIQUE as follows. First we introduce the notion of two literals L and L' being complementary. This means that either L is a positive literal x and L' is the corresponding negative literal $\neg x$ (with the same x), or vice versa. Given an instance $\varphi = (L_{11} \vee L_{12} \vee L_{13}) \wedge \ldots \wedge (L_{n1} \vee L_{n2} \vee L_{n3})$ of 3SAT, we transform it to an instance of CLIQUE by considering graph (V, E) with $3n$ nodes that stand for the $3n$ literal positions in φ and edges between any two nodes representing non-complementary literals within different clauses, and integer bound n. We show that formula φ is satisfiable if and only if (V, E) has a clique of size at least n. In one direction, if we have a truth-value assignment $\Im$ that satisfies φ we may choose from each clause of φ a literal that is satisfied by $\Im$. The chosen literals thus are non-complementary, hence the corresponding nodes form a clique of size n. In the other direction, having a clique of size n it must consist of nodes that represent n non-complementary literals taken from different clauses. Such a literal set is obviously satisfiable, thus φ is satisfiable. $\qquad\square$

A further, frequently used graph problem is VERTEX COVER. For an undirected graph (V, E), a subset C of V is called a *vertex cover* in case that for every edge in E at least one of its adjacent nodes belongs to C. Vertex cover problem VERTEX COVER is the problem of finding a vertex cover C of minimal size for an undirected graph (V, E). Stated as a decision problem it is to be answered whether an undirected graph (V, E) has a vertex cover of size at most b, for a given integer bound b.

VERTEX COVER
Given an undirected graph (V, E) and upper bound b, is there a vertex cover of size at most b.

Theorem 5.8.
VERTEX COVER is NP-complete.

Proof. 3SAT is polynomially reduced to VERTEX COVER as follows. Given instance $\varphi = (L_{11} \vee L_{12} \vee L_{13}) \wedge \ldots \wedge (L_{n1} \vee L_{n2} \vee L_{n3})$ of 3SAT containing Boolean variables $x_1, \ldots, x_m$, we transform it to an instance of VERTEX COVER by considering graph (V, E) with a triangle sub-graph for every clause, and a line sub-graph for every Boolean variable, and integer bound $2n + m$ as shown in Fig. 5.4.

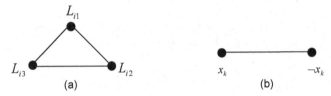

Fig. 5.4. (a) triangle representing i^{th} clause; **(b)** line representing k^{th} variable

Besides inner edges that occur within triangles and lines there are also edges between every triangle node and every line node that carry identical labels, that is, there is an edge from node labelled x_k to every node labelled L_{ij} with $L_{ij} = x_k$, and the same for $\neg x_k$. We show that formula φ is satisfiable if and only if (V, E) has a vertex cover of size at most $2n + m$. In one direction, let $\Im$ be a truth-value assignment that satisfies φ. From each line put that node into node set VC whose label evaluates 'true' in $\Im$. So far, every line edge is covered by node set VC. Furthermore, all edges between line nodes that are set to 'true' and corresponding triangle nodes are covered by node set VC. To also cover all inner triangle edges as well as all edges between line nodes that are set to 'false' and corresponding triangle nodes we simply put all triangle nodes into VC with the exception of a single, arbitrarily chosen literal that evaluates 'true' in $\Im$. Having two nodes from each triangle within VC, all inner triangle edges are covered by VC. As all triangle nodes that evaluate 'false' under $\Im$ are put into VC, all edges between line nodes set to 'false' and corresponding triangle nodes are covered by VC, too. Since every clause contains at most two literals that evaluate 'false' under $\Im$, VC is a vertex cover of size at most $2n + m$.

In the other direction, assume that VC is a vertex cover of size at most $2n + m$. Since line edges must be covered by node set VC, there must be at least one node from each line within VC. Thus, VC contains at least m line nodes. Since inner triangle edges must be covered by node set VC, there must be at least two nodes from each triangle within VC. Thus, VC contains at least $2n$ triangle nodes. Having at most $2n + m$ nodes, VC must contain exactly m line nodes (one from each line) and exactly $2n$ triangle nodes (two from each triangle). This opens way to define a truth-value assignment $\Im$ that evaluates variable x_k 'true' if and only if line node labelled x_k belongs to node set VC. For a triangle consider its single node u that does not belong to VC with literal x_k resp. $\neg x_k$ as label. The corresponding line node v with label x_k resp. $\neg x_k$ must be element of VC since otherwise we had an uncovered edge between u and v. This means that $\Im$ satisfies the literal associated with node u, thus $\Im$ also satisfies the clause represented by the triangle with node u. Thus $\Im$ satisfies all clauses. □

Next we deal with a couple of number problems. To obtain NP-completeness results requires some efforts since we must reduce from strongly structured domains (Boolean logic or graphs) to weakly structured domains of numbers. Our first problem is KNAPSACK which asks for finding a selection from a given list of numbers $w_1, \ldots, w_n$ that exactly sums to a given value W. Alternatively, it may be represented by the language of all number lists $w_1, \ldots, w_n, W$ such that a sub-list of $w_1, \ldots, w_n$ sums up to value exactly W.

KNAPSACK

Given list of numbers $w_1, \ldots, w_n, W$, is there a sub-collection of numbers $w_1, \ldots, w_n$ which sums up to exactly W.

Theorem 5.9.
3SAT is polynomially reducible to KNAPSACK, thus KNAPSACK is NP-complete.

Proof. 3SAT is polynomially reduced to KNAPSACK as follows. Given instance $\varphi = (L_{11} \vee L_{12} \vee L_{13}) \wedge \ldots \wedge (L_{n1} \vee L_{n2} \vee L_{n3})$ of 3SAT containing Boolean variables $x_1, \ldots, x_m$, we transform it to an instance of KNAPSACK by considering decimal numbers

$$a_1, \ldots, a_m, b_1, \ldots, b_m, c_1, \ldots, c_n, d_1, \ldots, d_n, G$$

each having $n + m$ digits as indicated in Fig. 5.5.

- The first n digits $a_{i1}a_{i2}\ldots a_{in}$ of a_i encode which of the clauses contains positive literal x_i by defining $a_{ij} = 1$ if clause $(L_{j1} \vee L_{j2} \vee L_{j3})$ contains x_i, and $a_{ij} = 0$ otherwise.
- The first n digits $b_{i1}b_{i2}\ldots b_{in}$ of b_i encode which of the clauses contains negative literal $\neg x_i$ by defining $b_{ij} = 1$ if clause $(L_{j1} \vee L_{j2} \vee L_{j3})$ contains $\neg x_i$, and $b_{ij} = 0$ otherwise.

- The other areas of the table form diagonal matrices with decimals 1 or 2 in the diagonals, or matrices consisting of decimals 0 only.

Note that j^{th} column contains between one and three digits 1 representing literals L_{j1}, L_{j2}, L_{j3} of the jth clause, a further digit 1 in line c_j and a further digit 2 in line d_j.

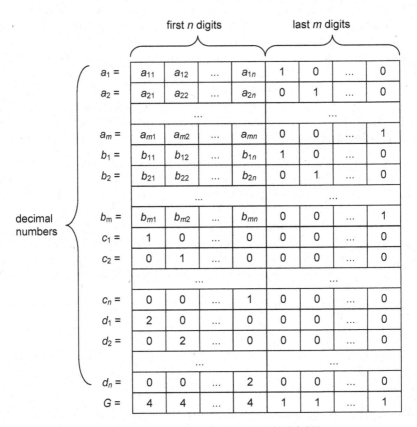

		first n digits				last m digits		
$a_1 =$	a_{11}	a_{12}	...	a_{1n}	1	0	...	0
$a_2 =$	a_{21}	a_{22}	...	a_{2n}	0	1	...	0
	...				...			
$a_m =$	a_{m1}	a_{m2}	...	a_{mn}	0	0	...	1
$b_1 =$	b_{11}	b_{12}	...	b_{1n}	1	0	...	0
$b_2 =$	b_{21}	b_{22}	...	b_{2n}	0	1	...	0
	...				...			
$b_m =$	b_{m1}	b_{m2}	...	b_{mn}	0	0	...	1
$c_1 =$	1	0	...	0	0	0	...	0
$c_2 =$	0	1	...	0	0	0	...	0
	...				...			
$c_n =$	0	0	...	1	0	0	...	0
$d_1 =$	2	0	...	0	0	0	...	0
$d_2 =$	0	2	...	0	0	0	...	0
	...				...			
$d_n =$	0	0	...	2	0	0	...	0
$G =$	4	4	...	4	1	1	...	1

decimal numbers {

Fig. 5.5. Reducing 3SAT to KNAPSACK

Now assume that formula φ is satisfied by a truth-value assignment $\Im$. Sum up all numbers a_i such that $\Im(x_i) =$ 'true' as well as all numbers b_i such that $\Im(x_i) =$ 'false'. Thus the first n decimals of the sum constructed so far range between 1 and 3 and hence can be suitably supplemented by additional value 1 and/or additional value 2 to give final decimals 4. The last m decimals of the sum constructed so far are 1's since exactly one of a_i or b_i was taken into the sum. This shows that final sum $G = 44 \ldots 411 \ldots 1$ can be achieved by summing up a suitable selection of the numbers in the table. Conversely, given a selection of numbers that sum up to exactly G, the diagonal unit matrices enforce that exactly one of a_i or b_i must have been taken into the sum. Thus,

$\Im(x_i) =$ 'true' if a_i was summed up, and $\Im(x_i) =$ 'false' if b_i was summed up, defines a truth-value assignment. As decimal 4 in j^{th} column can only be achieved by supplementary 1 and/or 2 in case that at least one a_i with $a_{ij} = 1$ or at least one b_i with $b_{ij} = 1$ was summed up, this means that clause $(L_{j1} \vee L_{j2} \vee L_{j3})$ contains at least one x_i with $\Im(x_i) =$ 'true' or at least one $\neg x_i$ with $\Im(x_i) =$ 'false', thus clause $(L_{j1} \vee L_{j2} \vee L_{j3})$ is satisfied by $\Im$.   □

Our second number problem is PARTITION. Here, a list of numbers $w_1, \ldots, w_n$ is to be partitioned into two non-empty sub-lists with identical sums. Formulated as a decision problem, it is asked whether a list of numbers $w_1, \ldots, w_n$ can be partitioned into two non-empty sub-lists with identical sums.

PARTITION
Given list of numbers $w_1, \ldots, w_n$, can it be partitioned into two sub-collections having identical sums.

Theorem 5.10.
KNAPSACK is polynomially reducible to PARTITION, thus PARTITION is NP-complete, too.

Proof. Let $w_1, \ldots, w_n, G$ be an instance of KNAPSACK. For the case that $w_1 + \ldots + w_n < G$ (instance without a solution) reduce this instance to the list consisting of numbers 1 and 2 as an instance (without solution) of PARTITION. In case that $w_1 + \ldots + w_n \geq G$ reduce this instance to list consisting of numbers $w_1, \ldots, w_n, a, b$ as an instance of PARTITION, with two extra numbers $a = G + 1$ and $b = w_1 + \ldots + w_n + 1 - G$. Assume that I is a subset of index set $\{1, \ldots, n\}$ such that $\sum_{i \in I} w_i = G$. Then compute

$$\sum_{i \in I} w_i + b = G + b = \sum_{i=1}^{n} w_i + 1$$

$$\sum_{i \notin I} w_i + a = \sum_{i=1}^{n} w_i - \sum_{i \in I} w_i + a = \sum_{i=1}^{n} w_i - G + a = \sum_{i=1}^{n} w_i + 1 \, .$$

Conversely assume that $w_1, \ldots, w_n, a, b$ is partitioned into two sub-lists with identical sums. Since $a + b = w_1 + \ldots + w_n + 2$ we know that the extra numbers a, b must be located within different sub-lists. Hence there is an index subset $I \subseteq \{1, \ldots, n\}$ such that

$$\sum_{i \in I} w_i + b = \sum_{i \notin I} w_i + a = \sum_{i=1}^{n} w_i - \sum_{i \in I} w_i + a \, .$$

Introducing the definitions of a and b this means that

$$2 \sum_{i \in I} w_i = \sum_{i=1}^{n} w_i + a - b = 2G \, .$$

Thus we obtain

$$\sum_{i \in I} w_i = G \ .$$

□

Coming back to graph problems we treat, as an example of a more complicated reduction, the problem HAMILTONIAN-PATH which is defined as follows: given an undirected graph with two distinguished nodes, start node x and goal node y, find out whether there exists a path starting at x and ending at y that visits each node in V exactly once (a so-called Hamiltonian path).

HP

Given an undirected graph (V, E) and two nodes x, y, decide whether there exists a Hamiltonian path from x to y.

Theorem 5.11.
3SAT is polynomially reducible to HP, thus HP is NP-complete, too.

Proof. Given a formula $\varphi = (L_{11} \vee L_{12} \vee L_{13}) \wedge \ldots \wedge (L_{n1} \vee L_{n2} \vee L_{n3})$ containing Boolean variables $x_1, \ldots, x_m$, we use a triangle sub-graph for every clause and a split sub-graph for every variable in the construction of graph (V, E). These are shown in Fig. 5.6. Nodes are either black or white. Black nodes will be, in addition to the links that are drawn in the diagrams below, completely connected, thus form a clique. In order not to overload graphs, these clique edges are not explicitly drawn. Note that colouring nodes and labelling edges serve the only purpose to simplify the description of the construction; colours and labels are not part of the graph, of course. Also keep in mind that later on, the "double edge" occurring at the moment in a split will vanish.

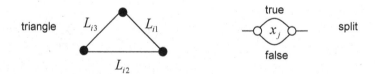

Fig. 5.6. Triangle and split subgraphs used in the reduction of 3SAT to HP

In the complete graph, splits $x_1, \ldots, x_m$ will be consecutively connected via their dangling edges. A further white start node is connected to the first split, the last split is connected to a further black node. Finally there is a further black goal node. Thus the arrangement of splits and start and goal nodes is presented in Fig. 5.7.

So far we have defined triangles and splits without any connections between them. Connections between triangles and splits are established by so-called

Fig. 5.7. Connecting split subgraphs

bridges which are defined to be sub-graphs with 4 "dangling edges" that receive target nodes later on in the construction of the complete graph. Bridges are used since they exhibit the following "zigzag" property: whenever in a Hamiltonian path a bridge is entered at one of its dangling edges, its nodes must necessarily be traversed in a "zigzag" manner as shown in Fig. 5.8 thus exiting the bridge on the same side as it was entered. The reader may try out some other of the numerous possible paths through a bridge and note that in all these cases some unvisited nodes are left that cannot be visited by a Hamiltonian path at a later time.

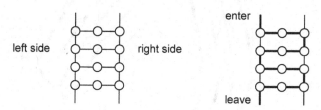

Fig. 5.8. "Zigzag" traversal through bridge subgraph

The graph that we are going to construct contains a bridge between each triangle edge labelled with positive literal x_j and the 'true' edge of split labelled x_j, and each triangle edge labelled with negative literal $\neg x_j$ and the 'false' edge of split labelled x_j. There are a few peculiarities of the connections to be clarified. The two dangling edges on one side of a bridge are connected to the nodes of the triangle edge that the bridge is connected with. Thus, every triangle edge is connected to exactly one split edge by a bridge. Split edges may be connected to more than one triangle edge in case that a literal has more than one occurrence in formula φ. In that case, dangling edges of consecutive bridges connected to the same split edge are melted into a single edge. One dangling edge of the first bridge is connected with the entry node of the split, one dangling edge of the last bridge is connected with the exit node of the split. Situation with two bridges thus looks as shown in Fig. 5.9. Also note that some split edge may not be connected to any triangle edge. This happens in case that either x_j or $\neg x_j$ has no occurrence in formula φ. Nevertheless, for every split, at least one of its edges is connected to a triangle edge by a bridge, since otherwise neither x_j nor $\neg x_j$ would occur in formula φ. This guarantees that splits do not possess multiple edges between its nodes.

In Fig. 5.9, imagine that a Hamiltonian path enters the left node of the split. The only way to visit all white nodes is traversing the first bridge in a zigzag manner, then entering the second bridge and traversing it in a zigzag manner, too, thus finally exiting at the right node of the split. Note that we have visited so far all white nodes, but no black node along this path. This brings us to the role of the black nodes. Any two of them are connected by an edge, thus black nodes form a clique. This opens way to visit black nodes and bridges not traversed so far.

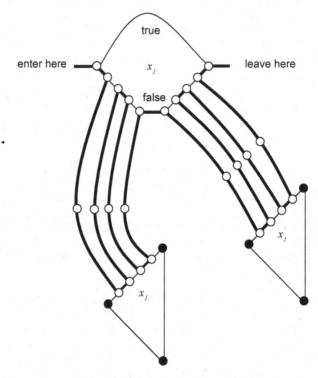

Fig. 5.9. Connecting splits to triangles via bridges with unique "zigzag" traversal

Lemma 5.12.
If formula φ is satisfiable the graph constructed above admits a Hamiltonian path from its start to its goal node.

Proof. Assume that truth-value assignment $\mathfrak{I}$ satisfies at least one literal within each clause of formula φ. Starting with the initial node traverse the arrangement of splits, using edge 'true' of split x_j in case that $\mathfrak{I}$ assigns value 'true' to x_j, and edge 'false' otherwise. Take care that on this traversal through the arrangement of splits every adjacent bridge is traversed in a zigzag manner. Note that for every triangle at least one of its adjacent bridges has been traversed since $\mathfrak{I}$ satisfied at least one literal within each clause. This traversal ends with the rightmost black node at the end of the arrangements of splits.

Up to now, all of the three black nodes of each triangle, as well as at most two adjacent bridges of each triangle are still unvisited. Since black nodes form a clique this allows us to visit all so far unvisited nodes as shown in Fig. 5.10, jumping from triangle to triangle and traversing so far unvisited bridges in the well-known "zigzag" manner.   □

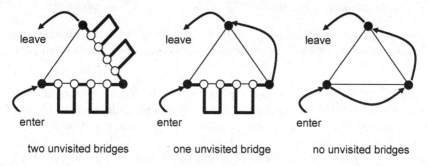

two unvisited bridges   one unvisited bridge   no unvisited bridges

Fig. 5.10. Visiting so far unvisited triangle nodes and bridges, depending on the number of so far unvisited adjacent bridges

Lemma 5.13.
Assume that the graph constructed above admits a Hamiltonian path from start to goal node. Then a truth-value assignment can be constructed that satisfies formula φ.

Proof. A Hamiltonian path starting at the initial node must necessarily traverse splits $x_1, \ldots, x_m$ in that order, for each split using either its 'true' or its 'false' path, and on this way must traverse all bridges that are connected to the traversed split edges (this is enforced by the zigzag property of bridges). Having arrived at the first black node at the end of the split arrangement, the remaining Hamiltonian path must visit all nodes belonging to triangles and all nodes of so far unvisited bridges. It can be easily seen that a Hamiltonian path cannot visit all nodes of a triangle if there are still three unvisited adjacent bridges. So, every triangle must have an adjacent bridge that has been already visited during the traversal through the arrangement of splits. Setting $\Im(x_j) = $ 'true' in case that the Hamiltonian path used the 'true' path in split x_j, and $\Im(x_j) = $ 'false' otherwise, thus defines a truth-value assignment that satisfies at least one literal (one that corresponds to an edge adjacent to an already visited bridge) in each clause.   □

With Lemma 5.12 and 5.13 the proof of Theorem 5.11 is complete.   □

The Hamiltonian path problem also arises for directed graphs. We refer to this as directed Hamiltonian path problem, DHP.

> **DHP**
> Given directed graph (V, E) and two nodes x, y,
> decide whether there exists a Hamiltonian path
> from x to y.

Theorem 5.14.
HP is polynomially reducible to DHP, thus DHP is NP-complete, too.

Proof. Replace an undirected graph by a directed one with two directed edges
(with opposite directions) for every undirected edge of the undirected graph.
$\square$

As a final NP-complete problem that we use in a bioinformatics reduction we
treat MAX-CUT which is defined as follows: given an undirected graph (V, E)
and an assignment of non-negative numbers to its edges, partition nodes into
two disjoint and non-empty sets A and B such that the sum of numbers
assigned to edges between nodes in A and nodes in B is a big as possible.

> **MAX-CUT**
> Given undirected graph (V, E) with weights as-
> signed to edges, and a lower bound b, decide
> whether there is a partition of V into subsets A, B
> with at least b edges between nodes of A and nodes
> of B.

As a preparation, we consider the following problem 4NAESAT with 'NAE'
standing for "not all equal". Problem 4NAESAT has as instances formulas of
the form $(L_{11} \vee L_{12} \vee L_{13} \vee L_{14}) \wedge \ldots \wedge (L_{n1} \vee L_{n2} \vee L_{n3} \vee L_{n4})$ with exactly
four literals per clause. It is to be answered whether a truth-value assignment
exists that makes at least one literal, but not all literals, within every clause
'true'.

> **4NAESAT**
> Given a Boolean formula in 4-clausal form, is there
> a truth-value assignment that makes at least one
> literal per clause 'true', and at least one literal per
> clause 'false'.

Theorem 5.15.
*3SAT is polynomially reducible to 4NAESAT, thus 4NAESAT is NP-complete,
too.*

Proof. Let an instance $\varphi = (L_{11} \vee L_{12} \vee L_{13}) \wedge \ldots \wedge (L_{n1} \vee L_{n2} \vee L_{n3})$ of 3SAT
be given. Expand it to an instance of 4NAESAT by introducing in each clause
the same new Boolean variable z obtaining formula

$$\psi = (L_{11} \vee L_{12} \vee L_{13} \vee z) \wedge \ldots \wedge (L_{n1} \vee L_{n2} \vee L_{n3} \vee z) .$$

Assume that truth-value assignment $\Im$ satisfies φ. Setting variable z to 'false' defines a truth-value assignment which satisfies between 1 and 3 literals within each clause of ψ . Conversely, let $\Im$ be a truth-value assignment that satisfies between 1 and 3 literals within each clause of ψ. Then the modified truth-value assignment $\neg\Im$ with $\neg\Im(x)$ = 'true' if $\Im(x)$ = 'false', and $\neg\Im(x)$ = 'false' if $\Im(x)$ = 'true' also satisfies between 1 and 3 literals per clause of ψ. Choose among $\Im$ and $\neg\Im$ the truth-value assignment which assigns 'false' to z. Obviously, this truth-value assignment satisfies every clause of φ.   □

It is known that the corresponding problem 3NAESAT for formulas with exactly three literals per clause is NP-complete, too. To show its NP-completeness is more complicated than for the problem 4NAESAT treated above. Since 4NAESAT is suitable to obtain NP-completeness of the last problem presented in this chapter, MAX-CUT, we have omitted a proof of the NP-completeness of 3NAESAT.

Theorem 5.16.
4NAESAT is polynomially reducible to MAX-CUT, thus MAX-CUT is NP-complete, too.

Proof. Let a 4SAT-formula

$$\varphi = (L_{11} \vee L_{12} \vee L_{13} \vee L_{14}) \wedge \ldots \wedge (L_{n1} \vee L_{n2} \vee L_{n3} \vee L_{n4})$$

containing Boolean variables $x_1, \ldots, x_m$ be given. Let a_j be the number of occurrences of x_j in φ, b_j the number of occurrences of $\neg x_j$ in φ, and n_j be $a_j + b_j$. Thus, $n_1 + \ldots + n_m = 4n$. Components of the graph we use are shown in Fig. 5.11. We are looking for a cut of size at least $6n$.

Fig. 5.11. Reducing 4NAESET to MAX-CUT

Lemma 5.17.
Assume that truth-value assignment $\Im$ satisfies between 1 and 3 literals within each clause. Then the weighted graph constructed above admits a cut of size at least $6n$.

Proof. Put into node set A of the cut all nodes that correspond to literals that are evaluated 'true' by $\Im$, and put into node set B of the cut all nodes that correspond to literals that are evaluated 'false' by $\mathfrak{J}$. Thus, edges between nodes that correspond to complementary literals contribute $n_1+\ldots+n_m = 4n$ to the size of the cut. Since every rectangle representing a clause has nodes in both parts of the cut, it contributes 2 or 4 to the size of the cut. (Question: Which distribution of nodes contributes 4 to the size of the cut?) Thus, all rectangles together contribute at least $2n$ to the size of the cut. All together, size of the cut is at least $6n$. □

Lemma 5.18.
Assume that the weighted graph constructed above admits a cut of size at least $6n$. Then there exists a truth-value assignment $\mathfrak{J}$ that satisfies between 1 and 3 literals within each clause of formula φ.

Proof. Consider a cut of size at least $6n$. If there is a variable x_j such that both nodes corresponding to x_j and $\neg x_j$ are within the same part of the cut, then these nodes contribute at most $2a_j + 2b_j = 2n_j$ to the size of the cut (remember that each node of a rectangle has exactly 2 adjacent edges that are weighted 1). By putting nodes corresponding to x_j and $\neg x_j$ into different parts of the cut we would gain n_j in size, but loose at most $2n_j$ in size, thus obtain a new cut with size that is still at least $6n$. Thus we may assume that for each variable x_j, nodes corresponding to x_j and $\neg x_j$ occur in different parts of the cut. Taken together, these nodes contribute exactly $4n$ to the size of the cut. Thus at least a value of $2n$ must be contributed to the size of the cut by rectangle edges. This only happens if each rectangle has nodes in both parts of the cut. Thus we have exactly the situation as in the proof of the lemma before. Defining truth-value assignment $\mathfrak{J}$ in such a way that exactly the literals corresponding to nodes in one part of the cut are set 'true' we enforce that in every clause between 1 and 3 literals are set 'true' by $\mathfrak{J}$. □

With Lemma 5.17 and 5.18 the proof of Theorem 5.16 is complete. □

5.4 Bridge to Bioinformatics: Shortest Common Super-Sequence

As an intermediate problem on the way towards NP-completeness of the sum-of-pairs multiple alignment problem MA, the shortest common super-sequence problem SSSEQ (see [53]) is shown to be NP-complete. This is done by reducing VERTEX COVER to SSSEQ. Later SSSEQ is further reduced to MA.

We fix some notations. We call string $T = T[1\ldots n]$ a *super-sequence* for string $S = S[1\ldots m]$ if the characters of S can be embedded under preservation of ordering into T, that is if there are indices $i_1,\ldots,i_m$ such that $1 \le i_1 < \ldots < i_m \le n$ and $S(1) = T(i_1),\ldots,S(m) = T(i_m)$. Thus the characters of S occur in a super-sequence T in the same order as they occur in S, though they need not occur contiguously in the super-sequence T. SSSEQ is the problem of finding a shortest common super-sequence for given strings

$S_1, \ldots, S_k$. Formulated as a decision problem, SSSEQ asks whether a common super-sequence T of length at most b exists, given strings $S_1, \ldots, S_k$ and upper bound b.

SSSEQ
Given strings $S_1, \ldots, S_k$ and upper bound b, is there a string T of length at most b that is a common super-sequence for $S_1, \ldots, S_k$.

Though SSSEQ may seem to be a not too complicated problem, it will turn out to be NP-complete. The following very simple example that is central to the reduction presented in this section gives some first indication that SSSEQ is more complex than one would expect at first sight. For a given number $k = 9n^2$ consider the following binary strings.

$$B_0 = 10^k \quad \ldots \quad B_i = 0^i 10^{k-i} \quad \ldots \quad B_k = 0^k 1$$

Thus B_i consists of a *left group* of i bits 0, followed by its single bit 1, followed by a *right group* of $k - i$ bits 0. It might seem that a shortest super-sequence for the collection of these strings $B_0, \ldots, B_k$ must be $0^k 10^k$ with the first block of k bits 0 serving the purpose of embedding the varying number of bits 0 of the left groups (in particular the k bits 0 of the left group of string B_k), and the last block of k bits 0 serving the purpose of embedding the varying number of bits 0 of the right groups (in particular the k bits 0 of the right group of string B_0). String $0^k 10^k$ consists of $2k + 1$ many bits. Astonishingly, there is a considerably shorter super-sequence. Consider the following string.

$$T = (0^{3n}1)^{3n}0^{3n}$$

String T is a super-sequence for the collection of strings $B_0, \ldots, B_k$, too. This can be seen as follows. Consider B_i. Embed the first i bits 0 of B_i into the first i bits 0 of T, and the single bit 1 in B_i to the next available bit 1 of T. By this, at most $3n$ bits 0 of T are omitted. Since T has $3n$ more bits 0 than B_i has, sufficiently many bits 0 are still available in T into which the last $k - i$ bits 0 of B_i may be embedded. Length of T is $9n^2 + 6n$, which is considerably shorter than $2k + 1 = 18n^2 + 1$. Interestingly, string T indeed is a shortest super-sequence for the collection of strings $B_0, \ldots, B_k$. To prove this requires some efforts.

Lemma 5.19.
Assume that R is a super-sequence for the collection of strings $B_0, \ldots, B_k$ defined above. Assume that R has x bits 1. Then R must possess at least $k - 1 + (k + 1)/x$ many bits 0. In particular, $(0^{3n}1)^{3n}0^{3n}$ is a shortest super-sequence for the collection of strings $B_0, \ldots, B_k$.

Proof. For simplicity assume that $k + 1$ is a multiple of x. (As an easy exercise, adapt the following proof to the case that $k + 1$ is not a multiple of x.) Embed each of the strings $B_0, \ldots, B_k$ into super-sequence R. As there are $k+1$ bits 1 mapped to bits 1 in R and R contains exactly x bits 1, there must be at least one bit 1 of R onto which $(k+1)/x$ or more of the bits 1 of strings $B_0, \ldots, B_k$ are mapped. Fix such a bit 1 in R. Let i_{min} be the smallest index of a string B_j whose bit 1 is mapped onto that fixed bit 1 of R, and i_{max} be the greatest index of a string B_j whose bit 1 is mapped onto that fixed bit 1 of R (Fig. 5.12). By choice of the fixed bit 1 of R, we know that $i_{max} - i_{min} + 1 \geq (k+1)/x$.

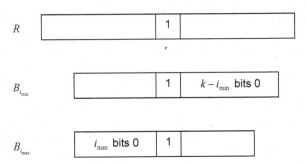

Fig. 5.12. First and last string whose bit 1 is mapped to the fixed bit 1 of string R

Furthermore, left of the fixed bit 1 in R there must exist at least i_{max} many bits 0 serving as targets for the i_{max} many bits 0 occurring left of bit 1 in $B_{i_{max}}$. Likewise, right of the fixed bit 1 in R there must exist at least $k - i_{min}$ many bits 0 serving as targets for the $k - i_{min}$ many bits 0 occurring right of bit 1 in $B_{i_{min}}$. Thus, the number of bits 0 in string R is at least $k + i_{max} - i_{min} \geq k - 1 + (k+1)/x$. Therefore the length of R is at least $x + k - 1 + (k+1)/x$. We prove $x + k - 1 + (k+1)/x \geq 9n^2 + 6n$ thus showing that $(0^{3n}1)^{3n}0^{3n}$ is a shortest super-sequence.

$$x + k - 1 + \frac{k+1}{x} \geq 9n^2 + 6n \Leftrightarrow x + 9n^2 - 1 + \frac{9n^2 + 1}{x} \geq 9n^2 + 6n$$

$$\Leftrightarrow x - 1 + \frac{9n^2 + 1}{x} \geq 6n \Leftrightarrow x + \frac{9n^2 + 1}{x} > 6n \Leftrightarrow x^2 + 9n^2 + 1 > 6nx$$

$$\Leftrightarrow x^2 - 6nx + 9n^2 + 1 > 0 \Leftrightarrow (x - 3n)^2 + 1 > 0$$

The last inequality is of course true.   □

Lemma 5.20.
Assume that R is a super-sequence for strings $B_0, \ldots, B_k$ defined above. Assume that R has $k - 1 + y$ bits 0. Then the number of bits 1 in R must be at least $(k+1)/y$.

Proof. Follows immediately from the previous lemma.   □

Theorem 5.21.
VERTEX COVER is polynomially reducible to SSSEQ. Thus SSSEQ is NP-complete.

Proof. The reduction runs as follows. As an instance of VERTEX COVER let be given an undirected graph (V, E) with $V = \{v_1, \ldots, v_n\}$, $E = \{e_1, \ldots, e_m\}$, and upper bound $k \leq n$ for the size of a vertex cover. Edges are always written as sets consisting of two nodes $e_l = \{v_a, v_b\}$ with $a < b$. We transform the given instance of VERTEX COVER into the instance of SSSEQ defined below. It consists of upper bound K for the length of a super-sequence, and strings A_iB_j and T_l, with $1 \leq i \leq n$, $1 \leq j \leq 9n^2$, $1 \leq l \leq m$ with $e_l = \{v_a, v_b\}$ and $a < b$ in the definition of string T_l below:

$$K = 15n^2 + 10n + k$$
$$A_i = 0^{6n(i-1)+3n}10^{6n(n+1-i)}$$
$$B_j = 0^j10^{9n^2-j}$$
$$T_l = 0^{6n(a-1)}10^{6n(b-a)-3n}10^{6n(n+2-b)}0^{9n^2}$$

For later use, we count the number of bits 0 and 1 in each of these strings. Each of strings A_i consists of $6n^2 + 3n$ bits 0 and a single bit 1, each of the strings B_j consists of $9n^2$ bits 0 and a single bit 1, thus each of strings A_iB_j and T_l consists of $15n^2 + 3n$ bits 0 and two bits 1.

First we show how a super-sequence of length K for the collection of strings A_iB_j and T_l is obtained from a vertex cover of size k.

Lemma 5.22.
Let a vertex cover $VC = \{v_{\sigma(1)}, \ldots, v_{\sigma(k)}\}$ of size k with $\sigma(1) < \ldots < \sigma(k)$ of graph (V, E) be given. Then the following string $S = S'S''$ is a super-sequence of length K for the collection of strings A_iB_j and T_l defined above. In the definition of S we repeatedly use substring $C = 0^{3n}10^{3n}$:

$$S' = C^{\sigma(1)-1}1C^{\sigma(2)-\sigma(1)}1\ldots1C^{\sigma(k)-\sigma(k-1)}1C^{n+1-\sigma(k)}0^{3n}$$
$$S'' = (0^{3n}1)^{3n}0^{3n}$$

Proof. We collect several properties of strings S' and S'' that are easily verified.

(1) S' consists of $6n^2 + 3n$ bits 0 and $n + k$ bits 1.
(2) S'' consists of $9n^2 + 3n$ bits 0 and $3n$ bits 1.
(3) S consists of $15n^2 + 6n$ bits 0 and $4n + k$ bits 1.
(4) S has length $K = 15n^2 + 10n + k$.
(5) Substring $C^{\sigma(d)-\sigma(d-1)}$ of S' consists of a left block 0^{3n}, a right block 0^{3n}, and several inner blocks $0^{3n}0^{3n}$, all of these blocks being separated by a bit 1. Consecutive substrings $C^{\sigma(d)-\sigma(d-1)}$ and $C^{\sigma(d+1)-\sigma(d)}$ are also separated by a bit 1.

$$S' = \underbrace{0^{3n}10^{6n}10^{6n}1\ldots10^{6n}10^{3n}}_{C^{\sigma(1)-1}}1\ldots$$

$$\underbrace{1\,0^{3n}10^{6n}10^{6n}1\ldots10^{6n}10^{3n}}_{C^{\sigma(d)-\sigma(d-1)}}\underbrace{1\,0^{3n}10^{6n}10^{6n}1\ldots10^{6n}10^{3n}}_{C^{\sigma(d+1)-\sigma(d)}}\ldots$$

(6) String S has exactly $3n$ more bits 0 than each of the strings A_iB_j and T_l has.

(7) The $(3n + 6n(i-1))^{\text{th}}$ bit 0 of S' is followed by a bit 1 (middle of some string C).

(8) If node v_i belongs to vertex cover VC, say $i = \sigma(d)$, then the $6n(i-1)^{\text{th}}$ bit 0 of S' is followed by a bit 1 (end of block $C^{\sigma(d)-\sigma(d-1)}$).

(9) If node v_i does not belong to vertex cover VC then the $6n(i-1)^{\text{th}}$ bit 0 of S' is followed by a block 0^{3n}.

□

Lemma 5.23.

String A_iB_j can be embedded into string S.

Proof. Figure 5.13 takes a closer look at the strings occurring in the lemma. First, walking into S' from its left end, prefix $0^{6n(i-1)+3n}$ of string A_iB_j is

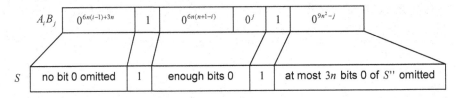

Fig. 5.13. Embedding A_iB_j into S

embedded into the first $2(i-1)+1$ blocks 0^{3n} of S'. Onto the then reached bit 1 in S' (see property (7) above), we map the left one of the two bits 1 of A_iB_j. Second, walking into S'' from its right end, suffix 0^{9n^2-j} of A_iB_j is embedded into the last $9n^2 - j$ bits 0 of S'', and the second of the two bits 1 in A_iB_j is mapped to the next available (walking further to the left) bit 1 in S''. Since j need not be a multiple of $3n$, mapping of this latter bit 1 eventually omits some bits 0 of S''. Nevertheless, the number of omitted bits 0 within S'' is obviously bounded by $3n$. Third, having so far mapped both bits 1 of A_iB_j onto bits 1 of S and omitted at most $3n$ bits 0 of S, there are still sufficiently many bits 0 available in the middle of string S for embedding the remaining bits 0 of A_iB_j (remember property (6) saying that S contains $3n$ more bits 0 than A_iB_j does).   □

Lemma 5.24.

String T_l with $l = \{v_a, v_b\}$ and $a < b$ can be embedded into string S.

Proof. First we consider the case that node v_a belongs to vertex cover VC, say $a = \sigma(d)$ for some d between 1 and k. Walking into S from its left end, prefix $0^{6n(a-1)}$ of T_l is embedded into the first $2(a-1)$ blocks 0^{3n} of S'. Onto the then arrived bit 1 immediately right of block $C^{\sigma(d)-\sigma(d-1)}$ we map the first of the two bits 1 of T_l. So far we have not omitted any bit 0 of S'. Walking further to the right in S, we next embed substring $0^{6n(b-a)-3n}$ of T_l into S. Either we immediately arrive at a bit 1 in S (Fig. 5.14 (a)) that is used to map the second of the two bits 1 of T_l, or after omitting $3n$ bits 0 of S we arrive at a bit 1 in S (Fig. 5.14 (b)) that is used to map the second of the two bits 1 of T_l. Having so far omitted at most $3n$ bits 0 of S, property (6) stated above tells us there are sufficiently many bits 0 left in S as target bits for the remaining bits 0 of T_l.

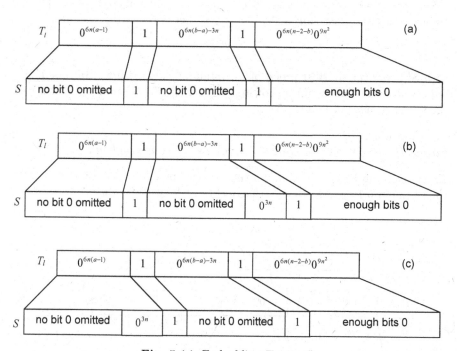

Fig. 5.14. Embedding T_l into S

Second we consider the case that node v_b belongs to vertex cover VC, say $b = \sigma(d)$ for some d between 1 and k. Walking into S from its left end, prefix $0^{6n(a-1)}$ of T_l is embedded into the first $2(a-1)$ blocks 0^{3n} of S'. Now we need not necessarily arrive at a bit 1 in S'. If we indeed did not arrive at a bit 1, omitting at most $3n$ bits 0 in S' leads us to a bit 1 onto which the first one of the two bits 1 of T_l can be mapped (Fig. 5.14 (c)). Walking further to the right in S, we next embed substring $0^{6n(b-a)-3n}$ of T_l into S. Here we arrive at the bit 1 immediately right of block $C^{\sigma(d)-\sigma(d-1)}$. This allows us to map the second of the two bits 1 of T_l onto this 1 of S' without omitting further bits 0 of S'. Having so far omitted at most $3n$ bits 0, property (6) stated above tells us there are sufficiently many bits 0 left in S as target bits for the remaining bits 0 of T_l. $\square$

Next we show how a vertex cover of size at most k is obtained from a super-sequence of length K for the collection of strings $A_i B_j$ and T_l.

Lemma 5.25.
Let S be a super-sequence of length $K = 15n^2 + 10n + k$ for the collection of strings $A_i B_j$ and T_l. From this a vertex cover $\{v_{\sigma(1)}, \ldots, v_{\sigma(q)}\}$ with $\sigma(1) <$ $\therefore \ldots < \sigma(q)$ of size $q \le k$ can be obtained.

Proof. Since each of the strings A_i contains $6n^2 + 3n$ bits 0, super-sequence S must contain at least $6n^2 + 3n$ bits 0, too. Let S be partitioned into

$$S = S'S''$$

with S' being the shortest prefix of S containing exactly $6n^2 + 3n$ bits 0. We require three further lemmas showing that string S' has similar structure as the string S' used in Lemma 5.22. $\square$

Lemma 5.26.
Strings B_j can be embedded into string S'', thus S'' has length at least $9n^2 + 6n$.

Proof. Looking at Fig. 5.15 shows that B_j can be embedded into string S'', thus Lemma 5.19 proves that S'' has length at least $9n^2 + 6n$. $\square$

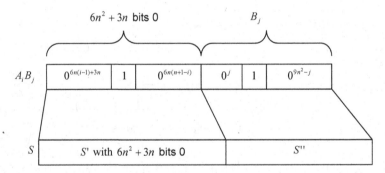

Fig. 5.15. B_j is embedded into S''

Lemma 5.27.
String S' contains between its $(6n(i-1) + 3n)^{\text{th}}$ and $6ni^{\text{th}}$ bit 0 at least one bit 1. In particular, S' contains at least n bits 1. (Compare this with property (7) from the proof of Lemma 5.22.)

Proof. Assume that S' does not contain any bit 1 between its $(6n(i-1)+3n)^{\text{th}}$ and $6ni^{\text{th}}$ bit 0. Since $0^{6n(i-1)+3n}1$ is a prefix of string A_i we conclude that the first bit 1 of A_i must be mapped onto some bit 1 right of the $6ni^{\text{th}}$ bit 0 of S'. Thus at least $3n$ bits 0 of S' are omitted when embedding A_i into S (Fig. 5.16). Now consider an arbitrary one of the strings B_j. Prefix string $0^{6n(i-1)+3n}10^{6n(n+1-i)}$ of A_iB_j contains $6n^2 + 3n$ bits 0, and string S' contains $6n^2 + 3n$ bits 0, too. Since at least $3n$ of bits 0 of S' are omitted when embedding A_i into S', we conclude that in an embedding of A_iB_j into S the remaining suffix $0^{3n}0^j10^{9n^2-j}$ of A_iB_j must be embedded into S''. Deleting the shortest prefix from S'' that contains exactly $3n$ bits 0 we obtain a string into which all of the strings $B_j = 0^j10^{9n^2-j}$ may be embedded. Since by Lemma 5.19 the shortest super-sequence for the collection of strings B_j has length $9n^2+6n$ we know that S'' has length at least $9n^2+9n$. Since S' contains by definition $6n^2 + 3n$ bits 0 we conclude that $S = S'S''$ has length at least $15n^2 + 12n$. This contradicts the fact that S has length $K = 15n^2 + 10n + k \leq 15n^2 + 10n + n = 15n^2 + 11n$. $\qquad\square$

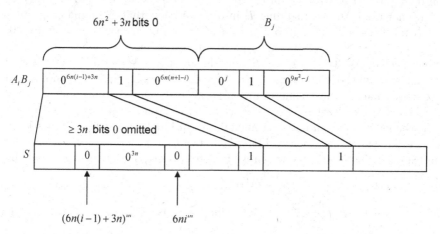

Fig. 5.16. Mapping the two bits 1 of A_iB_j to bits 1 of S

Lemma 5.28.
For all edges $e_l = \{v_a, v_b\}$ with $a < b$, string S' has at least one bit 1 either between its $6n(a-1)^{\text{th}}$ and $(6na - 3n)^{\text{th}}$ bit 0, or between its $6n(b-1)^{\text{th}}$ and $(6nb - 3n)^{\text{th}}$ bit 0. (Compare property (8) of the proof of Lemma 5.22 with this assertion.)

Proof. Assume that S' neither contains a bit 1 between its $6n(a-1)^{\text{th}}$ and $(6na - 3n)^{\text{th}}$ bit 0, nor between its $6n(b-1)^{\text{th}}$ and $(6nb - 3n)^{\text{th}}$ bit 0 (see Fig. 5.17). Thus, bit 1 in prefix $0^{6n(a-1)}1$ of string T_l must be mapped to some bit 1 of S right of the $(6na - 3n)^{\text{th}}$ bit 0. This means that in embedding prefix $0^{6n(a-1)}1$ of T_l we omit at least $3n$ bits 0 of S. Second bit 1 in T_l occurs

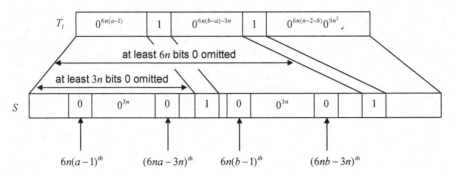

Fig. 5.17. Mapping the two bits 1 of T_l to bits 1 of S

immediately right of the $(6nb - 9n)^{\text{th}}$ bit 0. Since already $3n$ bits 0 of S have been omitted and right of the $6n(b-1)^{\text{th}}$ bit 0 of S' there are further $3n$ bits 0, we know that again at least $3n$ bits 0 of S' are omitted when embedding T_l into S. Since T_l has $15n^2 + 3n$ bits 0, S' has $6n^2 + 3n$ bits 0, and $6n$ bits 0 of S are not used when embedding T_l into S, we know that the remaining suffix of T_l must be embedded into S''. Thus S'' contains at least $6n + 9n^2$ bits 0. Denote the exact number of bits 0 in S'' by $p + 9n^2$ with some number $p \geq 6$. Since each of the strings A_i and S' contain the same number $6n^2 + 3n$ of bits 0 we know that all of the strings B_j must be embedded into S''. By Lemma 5.20 we know that S'' must contain at least $(9n^2 + 1)/(p + 1)$ many bits 1. Thus S'' has length at least $p + 9n^2 + (9n^2 + 1)/(p + 1)$. Since $K = 15n^2 + 10n + k$ was the length of S and S' contained $6n^2 + 3n$ bits 0 and at least n bits 1, we conversely know that the length of S'' can be at most $9n^2 + 6n + k$. This results in the following inequality which we lead to a contradiction (using $k \leq n$) by a sequence of transformation steps:

$$p + 9n^2 + \frac{9n^2 + 1}{p + 1} \leq 9n^2 + 6n + k \leq 9n^2 + 7n$$

$$\Leftrightarrow p + \frac{9n^2 + 1}{p + 1} \leq 7n . \qquad (i)$$

Now we show

$$6n + \frac{9n^2 + 1}{6n + 1} \leq p + \frac{9n^2 + 1}{p + 1} \qquad (ii)$$

by the following equivalence transformations.

$$6n + \frac{9n^2 + 1}{6n + 1} \le p + \frac{9n^2 + 1}{p + 1}$$

$$\Leftrightarrow p + \frac{9n^2 + 1}{p + 1} - 6n - \frac{9n^2 + 1}{6n + 1} \ge 0$$

$$\Leftrightarrow p(p + 1)(6n + 1) + (9n^2 + 1)(6n + 1)$$
$$- 6n(p + 1)(6n + 1) - (9n^2 + 1)(p + 1) \ge 0$$

$$\Leftrightarrow (p - 6n)(p + 1)(6n + 1) + (9n^2 + 1)(6n - p) \ge 0$$

$$\Leftrightarrow (p - 6n)((p + 1)(6n + 1) - 9n^2 - 1) \ge 0$$

$$\Leftrightarrow (p - 6n)(6np + 6n + p + 1 - 9n^2 - 1) \ge 0$$

The latter inequality is true since $p \ge 6n$ holds, thus also $6np \ge 9n^2$ holds within the second of the factors above. From (i) and (ii) we conclude:

$$6n + \frac{9n^2 + 1}{6n + 1} \le 7n$$

$$\Leftrightarrow \frac{9n^2 + 1}{6n + 1} \le n$$

$$\Leftrightarrow 9n^2 + 1 \le (6n + 1)n = 6n^2 + n$$

$$\Leftrightarrow 3n^2 \le n - 1 \,.$$

We arrive at a contradiction.   □

Remark 5.29.
The reader may point out why using the period string $C = 0^{3n}10^{3n}$ was essential and why, for example, $C = 0^{2n}10^{2n}$ or $C = 0^n10^n$ would not lead to a contradiction as above.

Now we can finish the construction of a vertex cover of size at most k. We collect some facts on S' and S'':

- S'' has length at least $9n^2 + 6n$ (Lemma 5.26).
- S has length $K = 15n^2 + 10n + k$ (assumption on S).
- S' has length at most $6n^2 + 4n + k$ (combine last two assertions).
- S' contains $6n^2 + 3n$ bits 0 (definition of S').
- S' has at most $n + k$ bits 1 (combine last two assertions).
- For all of the indices $1 \le i \le n$, between the $(6n(i - 1) + 3n)^{\text{th}}$ and $6ni^{\text{th}}$ bit 0 of S' there is at least one bit 1 (Lemma 5.27).
- For at most k of the indices $i = 1, \ldots, n$, there is a bit 1 within the interval between the $6n(i - 1)^{\text{th}}$ and the $(6n(i - 1) + 3n)^{\text{th}}$ bit 0 of S' (combine last two assertions). Let $\sigma(1) < \sigma(2) < \ldots < \sigma(q)$ with $q \le k$ be an enumeration of these intervals containing at least one bit 1.
- For every edge $e_l = \{v_a, v_b\}$ with $a < b$ at least one of indices a, b must belong to index set $\{\sigma(1), \ldots, \sigma(q)\}$ (combine last assertion and Lemma 5.28).
- Node set $\{v_{\sigma(1)}, \ldots, v_{\sigma(q)}\}$ is a vertex cover of (V, E) of size at most k.

                                                                □

5.5 NP-Completeness of Core Bioinformatics Problems

5.5.1 Multiple Alignment

The MULTIPLE ALIGNMENT problem under sum-of-pairs scoring (MA) is the problem of finding a multiple alignment $T_1, \ldots, T_k$ having *maximum* SP-score for given strings $S_1, \ldots, S_k$ and scoring function σ. Formulated as a decision problem it is to be decided whether a multiple alignment $T_1, \ldots, T_k$ with SP-score *at least* M exists for given strings $S_1, \ldots, S_k$, scoring function σ and lower bound M. Using the specific scoring function σ shown in Fig. 5.18 and strings over the 4-letter alphabet $\Sigma = \{0, 1, a, b\}$, we show that SSSEQ can be polynomially reduced to this specialization of SP-Align. Thus SP-Align with this specific scoring function, as well as SP-Align in general is NP-complete (see [77]). Values of the used scoring function σ are chosen in such a way that reduction of SSSEQ to MA and numerical computations are as easy as possible.

σ	0	1	a	b	-
0	-4	-4	-1	-2	-2
1	-4	-4	-2	-1	-2
a	-1	-2	0	$-\infty$	-2
b	-2	-1	$-\infty$	0	-2
-	-2	-2	-2	-2	0

Fig. 5.18. Scoring matrix used in the reduction of SSSEQ to MA

Consider now an instance of SSSEQ consisting of binary strings $S_1, \ldots, S_k$ and upper bound m for the length of a common super-sequence. Define

$$s = |S_1| + \ldots + |S_k| \ . \tag{5.1}$$

For both directions of the reduction defined below we require the following lemma.

Lemma 5.30.
Each multiple alignment $T_1, \ldots, T_k$ of $S_1, \ldots, S_k$ has the same SP-score $-2s(k-1)$.

Proof. Consider a fixed column of an alignment $T_1, \ldots, T_k$. Assume that it contains x bits 0 or 1, and $k - x$ spacing symbols. Comparisons between bits and spacing symbols contribute to SP-score the following value:

$$x(k-x)(-2) = -2xk + 2x^2 \ .$$

Comparisons among any two bits contribute to SP-score the following value:

$$\frac{1}{2}(x-1)x(-4) = -2x^2 + 2x .$$

Comparisons among any two spacing symbols contribute 0 to the SP-score. Thus, any column containing x bits contributes $-2x(k-1)$ to SP-score. Summarizing over all columns leads to SP-score $-2s(k-1)$ as s was defined to be the number of bits 0 and 1 occurring in $T_1, \ldots, T_k$.   □

We now *truth-table reduce* (see Sect. 5.1 for an explanation of this notion) the considered instance consisting of binary strings $S_1, \ldots, S_k$ and upper bound m to $m+1$ different instances of SP-Align, $\Im(i)$, for $i = 0, 1, \ldots, m$, and lower bound M as follows, with a^i, b^{m-i} denoting strings that consist of i resp. $m-i$ repetitions of characters 'a' resp. 'b':

$$\Im(i) = S_1, \ldots, S_k, a^i, b^{m-i}$$
$$M = -2s(k-1) - 3s - m .$$

Lemma 5.31.
Assume that strings $S_1, \ldots, S_k$ have a super-sequence T of length m consisting of i bits 0 and $j = m - i$ bits 1. Then from T a multiple alignment of strings $S_1, \ldots, S_k, a^i, b^{m-i}$ with SP-score M can be obtained.

Proof.

- Write the characters of S_p below identical characters of T corresponding to an embedding of S_p into super-sequence T, for $p = 1, \ldots, k$.
- Write the characters of string a^i below the i bits 0 of T.
- Write the characters of string b^{m-i} below the $m - i$ bits 1 of T.
- At all positions not filled so far with a character, write a spacing symbol.
- Call the resulting strings $T_1, \ldots, T_k, A, B$. These form a multiple alignment of $S_1, \ldots, S_k, a^i, b^{m-i}$.

As an example, Fig. 5.19 shows the multiple alignment constructed this way for strings $S_1 = 01011001$, $S_2 = 101111$, $S_3 = 00000011$, and super-sequence $T = 00010110111011$. Here, there are $i = 6$ bits 0, and $j = 8$ bits 1 in T.

We compute the score of the constructed multiple alignment. The contribution of $T_1, \ldots, T_k$ to SP-score was shown above to be $-2s(k-1)$. Since every 0 in $T_1, \ldots, T_k$ is aligned in A and B to 'a' and -, and every 1 is aligned to - and 'b', the contribution to SP-score that results from comparisons between all of $T_1, \ldots, T_k$ and all of A, B is $-s - 2s = -3s$. Finally, comparison of A and B leads to a further contribution of $-m$ to SP-score. Summarizing all contributions leads to a contribution of

$$M = -2s(k-1) - 3s - m .$$

□

T	0	0	0	1	0	1	1	0	1	1	1	0	1	1
T_1	-	0	-	1	0	1	1	0	-	-	-	0	1	-
T_2	-	-	-	1	0	1	-	-	1	1	-	-	1	-
T_3	0	0	0	-	0	-	-	0	-	-	-	0	1	1
A	a	a	a	-	a	-	-	a	-	-	-	a	-	-
B	-	-	-	b	-	b	b	-	b	b	b	-	b	b

Fig. 5.19. Example multiple alignment

Lemma 5.32.
Let i be any number between 0 and m. From a multiple alignment of strings $S_1, \ldots, S_k, \mathrm{a}^i, \mathrm{b}^{m-i}$ with SP-score $\Gamma \geq M$, a super-sequence T of length m for $S_1, \ldots, S_k$ can be obtained.

Proof. Let $T_1, \ldots, T_k, A, B$ be a multiple alignment of $S_1, \ldots, S_k, \mathrm{a}^i, \mathrm{b}^{m-i}$ with SP-score

$$\Gamma \geq M = -2s(k-1) - 3s - m .$$

Let n be the length of each of the strings $T_1, \ldots, T_k, A, B$. As was already computed in the lemma above, the contribution of strings $T_1, \ldots, T_k$ to Γ is

$$= -2s(k-1) .$$

In order for $\Gamma \geq M$ to hold, comparison of all of strings A, B to all of strings $T_1, \ldots, T_k$, as well as comparison of A with B must further contribute to Γ a value

$$\geq -3s - m .$$

As a consequence, there cannot be any alignment of 'a' with 'b' since this would contribute value $-\infty$ to the SP-score. Thus we conclude that $n \geq m$ holds. As A, B contain no other alignments than 'a' with -, - with 'b', and - with -, comparison of A with B contributes to SP-score the value

$$= -m .$$

This finally means that the contribution to SP-score which results from comparisons of all of strings $T_1, \ldots, T_k$ and all of A, B must be

$$\geq -3s .$$

Note that every alignment of a bit within one of $T_1, \ldots, T_k$ and a character or spacing symbol within A, B contributes a value as shown in Fig. 5.20 to the overall SP-score. In order for all contributions to sum up to some value $\geq -3s$, each of the s bits of $T_1, \ldots, T_k$ must contribute exactly -3 to Γ. Thus

bit in $T_1,\cdots,T_k$	character in A	character in B	contribution
0	a	-	-3
0	-	b	-4
0	-	-	-4
1	a	-	-4
1	-	b	-3
1	-	-	-4

Fig. 5.20. Contributions to SP-score

each bit 0 must be aligned with 'a' and -, and each bit 1 must be aligned with - and 'b'. As a consequence, there is no pairing of - with - in the alignment of A and B, thus $n \leq m$. We have now shown that the alignment exactly looks like the alignment constructed in the proof of the lemma before. Taking as string T the string that has bits 0 at all positions where string A has a character 'a', and bits 1 at all positions where string B has a character 'b', defines a string of length m into which every string S_i is embedded.   □

Thus we have shown the following theorem.

Theorem 5.33.
SSSEQ can be truth-table reduced in polynomial time to MA, thus MA cannot be solved in polynomial time (unless $P = NP$).

5.5.2 Shortest Common Superstring

The shortest common super-string problem SSSTR is the problem of finding a shortest string S that contains each of a given list of strings $S_1, \ldots, S_k$ as a sub-string. Formulated as a decision problem, SSSTR asks whether a superstring T of length at most b exists, given strings $S_1, \ldots, S_k$ and number b.

Theorem 5.34.
DIRECTED HAMILTONIAN PATH is polynomially reducible to SSSTR. Thus, SSSTR is NP-complete, too.

Proof. Let a directed graph with n nodes $v_1, \ldots, v_n$ and m edges be given with distinguished source node v_1 and distinguished target node v_n. Let edges be given in form of an adjacency list $L(v)$, for every node v. The following characters are used:

$\alpha, \beta, \#$ (initializing, finalizing, and separating character)
$v_1, \ldots, v_n$ (corresponding to the nodes of the considered graph)
$V_1, \ldots, V_n$ (copies of the characters introduced before)

Thus for every *node* v there are two *characters*, v and V. Now we consider the following strings each being of length 3: for every adjacency list $L(v) = [w_1, \ldots, w_k]$

$$\{V w_i V \mid i = 1, \ldots, k\} \cup \{w_i V w_{i+1} \mid i = 1, \ldots, k-1\} \cup \{w_k V w_1\}$$

(*"scroll terms"*)

as well as the following strings:

$$\{v \# V \mid v \in V, v \neq v_1 \text{ and } v \neq v_n\}$$

(*"coupling terms"*)

$$\alpha \# v_1$$

(*"initializing term"*)

$$v_n \# \beta$$

(*"finalizing term"*)

For every $i = 1, \ldots, k$, the following string is a shortest common super-string for the scroll terms associated with adjacency list $L(v) = [w_1, \ldots, w_k]$:

$$\text{scroll}(L(v), w_i) = V w_i V w_{i+1} V \ldots w_{k-1} V w_k V w_1 V w_2 \ldots V w_{i-1} V w_i$$

This is clear as always a maximum possible overlap of 2 characters between consecutive strings is achieved. Note that any string $\text{scroll}(L(v), w_i)$ contains twice as many characters as there are outgoing edges at node v, plus two further characters, as initial node w_i is repeated at the end.

As an example, consider node v with adjacency list $L(v) = [a, b, w, c, d]$ and node w occurring in $L(v)$ with adjacency list $L(w) = [e, f, g, h]$. Selecting node w as initial node in the first scroll string, and, for example, node e as initial node in the second scroll string, we obtain:

$$\text{scroll}(L(v), w) = V w V c V d V a V b V w$$
$$\text{scroll}(L(w), e) = W e W f W g W h W e.$$

Using coupling string $w \# W$, we may join these two strings in such a way that maximum overlap of one character left and one character right of $w \# W$ is achieved. Note that this maximum overlap between any two such scroll strings and coupling string is possible if and only if node w appears in $L(v)$, and we used $\text{scroll}(L(v), w)$ with w as start node.

$$\text{scroll}(L(v), w) \# \text{scroll}(L(w), e) = V w V c V d V a V b V w \# W e W f W g W h W e$$

Lemma 5.35.
Assume that $w_1, w_2, \ldots, w_n$ is a Hamiltonian path from node v_1 to node v_n in the considered graph. Then the following string S is a common super-string (even a shortest one, but this does not matter here) for the scroll, coupling, initializing, and finalizing terms associated with the graph.

$$S = \alpha\#scroll(L(w_1), w_2)\#scroll(L(w_2), w_3)\# \ldots \#scroll(L(w_{n-1}), w_n)\#\beta$$

It length is $2m + 3n$ (note that n was the number of nodes, and m was the number of edges).

Proof. *Scroll* terms appear as sub-strings within their associated *scroll* strings. Every *coupling* term is used in the string defined above, since a Hamiltonian path covers all nodes. *Initializing* and *finalizing* terms also appear at the beginning and end of the string. Within the $n-1$ *scroll* strings there are $2m$ characters (distributed among the different *scroll* strings) plus $2(n-1)$ characters as within every *scroll* string the initial node is repeated at the end. Right of any of the $n-1$ *scroll* strings there appears the separating character $\#$. On the left there are the characters $\alpha\#$ at the beginning and β at the end. All together, the constructed super-string has length

$$|S| = 2m + 2(n-1) + (n-1) + 3 = 2m + 3n \,.$$

$\square$

Lemma 5.36.
Assume that T is a common super-string of length at most $2m + 3n$ for the scroll, coupling, initializing, and finalizing terms associated with the graph. This constrains T to look exactly like the string S constructed in the lemma before. In particular, a Hamilton path $w_1, w_2, \ldots, w_n$ from node v_1 to node v_n can be extracted from T.

Proof. First two characters $\alpha\#$ of initializing term, last two characters $\#\beta$ of finalizing term, and $n-2$ characters $\#$ in coupling terms cannot occur in an overlapping manner within super-string T. Thus there are exactly $2m + 3n - 2 - 2 - n + 2 = 2m + 2n - 2 = 2m + 2(n-1)$ characters left for embedding as substrings all scroll terms. As all scroll terms together contain $2m + 2(n-1)$ characters this is possible only by using for every adjacency list $L(w_i)$ some string $scroll(L(w_i), w_{i+1})$ with w_{i+1} taken from $L(w_i)$ and w_{i+1} being the node that defines the next term $scroll(L(w_{i+1}), w_{i+2})$. This means that the consecution of used sub-strings $scroll(L(w_i), w_{i+1})$ defines a Hamiltonian path from node v_1 to node v_n. $\square$

With Lemma 5.36 and 5.37 the proof of Theorem 5.35 is complete. $\square$

Remark 5.37. In the reduction of DHP to SSSTR we used an alphabet that dynamically depends on the instance of DHP. Alternatively, we may always reduce to an instance of SSSTR consisting of strings over a fixed alphabet, for example $\{L, R, 0\}$. For this, simulate characters $a_1, \ldots, a_p$ by strings $L0R, L00R, \ldots, L0^p R$. The number of bits 0 encodes which one of the former characters is meant, brackets L and R prevent overlaps of non-zero length between different strings. Using binary alphabet would require more involved string encodings of characters.

5.5.3 Double Digest

Theorem 5.38.
PARTITION is polynomially reducible to DDP, thus DDP is NP-complete, too.

Proof. Here we will see an example of an almost trivial reduction showing how valuable a rich zoo of known NP-complete problems is, in particular if that zoo also contains rather simply defined problems coming from weakly structured domains (e.g. number problems). Consider an instance $w_1, \ldots, w_n$ of PARTITION. In case that $W = w_1 + \ldots + w_n$ is odd, hence the instance is an unsolvable instance of PARTITION, we reduce it to an arbitrary instance of DDP that is unsolvable, too. The reader may construct such an instance. In case that $W = w_1 + \ldots + w_n$ is even, we reduce to the following three lists:

$$A = w_1, \ldots, w_n \qquad B = \tfrac{1}{2}W, \tfrac{1}{2}W \qquad C = w_1, \ldots, w_n$$

Then permutations A^*, B^*, C^* of A, B, C exist such that the superimposition of A^* with B^* gives C^* if and only if $w_1, \ldots, w_n$ can be partitioned into two disjoint and non-empty sub-lists with identical sums (use that B^* contributes a single cut exactly at position $\tfrac{1}{2}M$). $\qquad\square$

5.5.4 Protein Threading

Given the primary structure of a new protein $S = S[1 \ldots n]$ and the loop-core structure of a known protein $T = L_0 C_1 L_1 C_2 \ldots C_m L_m$ with strongly conserved core segment lengths $c_1, \ldots, c_m$ and intervals $[\lambda_0, \Lambda_0], \ldots, [\lambda_m, \Lambda_m]$ restricting the loop segment lengths, let PT be the problem of computing the maximum value $f(t_1, \ldots, t_m)$ of a scoring function for threadings $t_1, \ldots, t_m$ of S into the loop-core structure of T of the form length

$$f(t_1, \ldots, t_m) = \sum_{i=1}^{m} g(t_i) + \sum_{i=1}^{m-1} \sum_{j=i+1}^{m} h(t_i, t_j) . \tag{5.2}$$

Theorem 5.39.
MAX-CUT is polynomially reducible to PT , thus PT is NP-complete, too.

Proof. Let an instance of MAX-CUT, that is set of nodes $V = \{v_1, \ldots, v_m\}$, set of undirected edges E, and a lower bound b for the desired size of a cut, be given. It is asked whether there is a cut $V = V_1 \cup V_2$ with at least b many edges of E between V_1 and V_2. We reduce this to the following instance of PT, where we mainly encode the task of maximizing the number of edges between the components of a cut into the coupling part

$$\sum_{i=1}^{m-1} \sum_{j=i+1}^{m} h(t_i, t_j)$$

of the scoring function. This is not surprising in sight of the fact that without this coupling term PT would be a polynomially solvable problem (as was shown in Chap. 3). Details of the reduction are as follows.

- $S = (01)^m$
- $c_1 = c_2 = \ldots = c_m = 1$
- $\lambda_0 = \lambda_1 = \ldots = \lambda_m = 0$ and $\Lambda_0 = \Lambda_1 = \ldots = \Lambda_m = \infty$
- $h(t_i, t_j) = \begin{cases} 1 & \text{if } \{v_i, v_j\} \in E \text{ and } S(t_i) \neq S(t_j) \\ 0 & \text{otherwise} \end{cases}$
- b as lower bound for the value of $\sum_{i=1}^{m-1} \sum_{j=i+1}^{m} h(t_i, t_j)$

Lemma 5.40.
Assume that parameters are fixed as above.
(a) From a cut $V = V_1 \cup V_2$ with at least b many edges from E between V_1 and V_2 we obtain a threading $t_1, \ldots, t_m$ with score at least b.
(b) From a threading $t_1, \ldots, t_m$ with score at least b we obtain a cut $V = V_1 \cup V_2$ with at least b many edges from E between V_1 and V_2.
Thus, (V, E) has a cut with at least b many edges between its parts if and only if the constructed instance of PT has a threading with score at least b. This is the desired polynomial reduction from MAX-CUT to PT.

Proof. (a) Consider cut $V = V_1 \cup V_2$ with at least b many edges from E between V_1 and V_2. To thread the ith core segment into $S = (01)^m$ consider node v_i. In case that $v_i \in V_1$ choose bit 0 of the ith substring (01) of S as start position for the ith core segment. In case that $v_i \in V_2$ choose bit 1 of the ith substring (01) of S as start position for the ith core segment. For each edge $\{v_i, v_j\}$ in E with nodes in different parts of the cut, that is with either $v_i \in V_1$ and $v_j \in V_2$, or with $v_i \in V_2$ and $v_j \in V_1$, we obtain a contribution of 1 to the scoring function as different bits are selected for $S(t_i)$ and $S(t_j)$. Thus the threading also has score at least b.
(b) Let a threading $t_1, \ldots, t_m$ with score at least b be given. By definition of the scoring function there must be at least b contributions 1. Contribution 1 occurs only for pairs t_i and t_j with an edge $\{v_i, v_j \in E\}$ and different bits $S(t_i)$ and $S(t_j)$. Defining $V_1 = \{v_i \mid S(t_i) = 0\}$ and $V_2 = \{v_i \mid S(t_i) = 1\}$ thus defines a cut with at least b edges between V_1 and V_2. □

With Lemma 5.41 the proof of Theorem 5.40 is complete. □

5.5.5 Bi-Clustering

Let (V, W, F) be a bipartite graph with node sets V and W, and set of edges $\{v, w\}$ between certain node pairs $v \in V$ and $w \in W$. A bi-clique consists of subsets $A \subseteq V$ and $B \subseteq W$ such that for all $a \in A$ and $b \in B$ there is an edge $\{a, b\}$ in F. Visualizing bipartite graphs and bi-cliques in a rectangular diagram (Fig. 5.21) makes clear that finding a bi-clique with maximum number of edges $|A|\,|B|$ is exactly the formal problem behind finding a maximum size sub-matrix (after permutations of rows and columns) of a microarray data matrix that consists of entries 1 only.

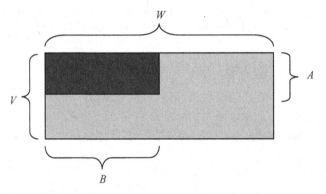

Fig. 5.21. A bi-clique

Lemma 5.41.
Given 3SAT-formula

$$\varphi = (L_{11} \vee L_{12} \vee L_{13}) \wedge (L_{21} \vee L_{22} \vee L_{23}) \wedge \ldots \wedge (L_{n1} \vee L_{n2} \vee L_{n3})$$

with n clauses $(L_{i1} \vee L_{i2} \vee L_{i3})$ each containing three literals L_{i1}, L_{i2}, L_{i3}, an undirected graph (V, E) consisting of $4n$ nodes can be constructed such that φ is satisfiable if and only if (V, E) has a clique of size exactly $2n$ (thus this is a reduction to the variant of CLIQUE where we ask for a clique having exactly half as many nodes as the graph has, called $\frac{1}{2}$-CLIQUE).

$\frac{1}{2}$-CLIQUE
Given undirected graph with $4n$ nodes, does it possess a clique with $2n$ nodes.

Proof. Take a look at the reduction of 3SAT to CLIQUE that was used in Theorem 5.7. There, a graph consisting of $3n$ nodes was used and the clique that occurred had exactly size n. Inserting further n nodes that are completely connected to all of the former $3n$ nodes establishes the assertion of the lemma. □

Theorem 5.42.
$\frac{1}{2}$-*CLIQUE is polynomially reducible to BI-CLIQUE, thus BI-CLIQUE is NP-complete, too.*

Proof. Given an undirected graph (V, E) with node set $V = \{v_1, \ldots, v_n\}$ of size $n = 2k \geq 16$ with even number k and edge set $E = \{e_1, \ldots, e_m\}$, it can be transformed into a bipartite graph (V, W, F) (first node set of the bipartite graph is indeed the same as the node set from (V, E)) such that (V, E) has a clique C with $|C| \geq k$ if and only if (V, W, F) has a bi-clique consisting of subsets A and B such that $|A|\,|B| \geq k^3 - \frac{1}{2}3k^2$ (note that $k \geq 3$ and k is an even number).

We define node set W and edge set F as follows using (somehow unusual, nevertheless admissible) the edges of E as nodes in the second component W together with a number of $\frac{1}{2}k^2 - k$ fresh nodes in W.

$$V = \{v_1, \ldots, v_n\}$$
$$W = \left\{e_1, \ldots, e_m, f_1, \ldots, f_{\frac{1}{2}k^2 - k}\right\}$$
$$F = \{\{v_i, e_j\} \mid 1 \leq i \leq n, 1 \leq j \leq m, v_i \notin e_j\}$$
$$\cup \left\{\{v_i, f_j\} \mid 1 \leq i \leq n, 1 \leq j \leq \frac{1}{2}k^2 - k\right\}$$

We show that the desired reduction property holds. In one direction, assume that (V, E) has a clique of size at least k. Take a clique C of size exactly k. Define bi-clique A, B as follows:

$$A = V - C$$
$$B = \left\{f_1, \ldots, f_{\frac{1}{2}k^2 - k}\right\} \cup \{e_j \mid 1 \leq j \leq m, e_j \subseteq C\}\ .$$

Thus, B contains all fresh nodes as well as all edges from E that connect nodes of clique C. As C is a clique, there is an edge $\{a, b\}$ in E for any two different nodes a, b in C. Size of the defined bi-clique is thus calculated as follows:

$$|A| = \frac{k}{2} = |C|$$
$$|B| = \frac{1}{2}k^2 - k + \frac{1}{2}k(k - 1)$$
$$|A|\,|B| = k\left(\frac{1}{2}k^2 - k + \frac{1}{2}k(k-1)\right) = k^3 - \frac{1}{3}k^2\ .$$

In the converse direction, assume that (V, W, F) has a bi-clique consisting of subsets $A \subseteq V$ and $B \subseteq W$ such that $|A|\,|B| \geq k^3 - \frac{3}{2}k^2$ holds. We may assume that B contains all of the fresh nodes (otherwise simply put the missing nodes into B, obtaining again a bi-clique of even larger size) as well as b many of the edges from E as nodes. By definition of edge set F we know that edges e_j

occurring in B do not contain any nodes from A. Thus they consist of nodes from $V - A$ only. By renumbering edges in E we may assume without loss of generality that B looks as follows:

$$B = \left\{ f_1, \ldots, f_{\frac{1}{2}k^2 - k} \right\} \cup \{ e_j \mid 1 \leq j \leq b \} \text{ with } e_j \subseteq (V - A) \times (V - A) .$$

Now consider node set $C = V - A$. Define $a = |A|$. Thus, the considered bi-clique is located as shown in Fig. 5.22. As C consists of $2k - a$ many nodes, the number of edges connecting nodes from C, and thus also number b can be bounded as follows:

$$b \leq \frac{1}{2}(2k - a)(2k - a - 1) .$$

We show that $a \leq k$ follows from this. Assume that $a > k$ holds. It is convenient to further compute with $x = a - k$. By definition of x and from $a \leq 2k$ we conclude that $0 < x \leq k$ holds. Furthermore, $2k - a = k - x$ holds. We obtain a contradiction as follows (the last estimation uses $x \leq k$):

$$0 \leq |A| \, |B| - k^3 + \frac{3}{2}k^2$$

$$\leq (k + x)\left(\frac{1}{2}k^2 - k + \frac{1}{2}(k - x)(k - x - 1) \right) - k^3 + \frac{3}{2}k^2$$

$$= \frac{1}{2}k^3 - k^2 + \frac{1}{2}xk^2 - xk + \frac{1}{2}\left(k^2 - x^2 \right)(k - x - 1) - k^3 + \frac{3}{2}k^2$$

$$= \frac{1}{2}k^3 - k^2 + \frac{1}{2}xk^2 - xk + \frac{1}{2}k^3 - \frac{1}{2}xk^2 - \frac{1}{2}k^2 - \frac{1}{2}x^2k + \frac{1}{2}x^3 + \frac{1}{2}x^2$$

$$- k^3 + \frac{3}{2}k^2$$

$$= -xk - \frac{1}{2}x^2k + \frac{1}{2}x^3 + \frac{1}{2}x^2$$

$$= \frac{1}{2}x(x^2 - xk + x - 2k)$$

$$= \frac{1}{2}x(x^2 - x(k - 1) - 2k)$$

$$< 0 .$$

So far we know that $a \leq k$ holds. In the following it is convenient to compute with $y = k - a$. By definition of y we conclude that $0 \leq y \leq k$. Now we argue by contradiction. Assume that (V, E) does not contain a clique of size k. Using that C has $2k - a = k + y$ many elements, we may successively select nodes pairs $(p_1, q_1), \ldots, (p_{y+1}, q_{y+1})$ such that the following properties holds (remember the assumption that there is no clique of size k):

- $p_1, q_1 \in C, p_1 \neq q_1, \{p_1, q_1\} \notin E$
- $p_2, q_2 \in C - \{p_1\}, p_2 \neq q_2, \{p_2, q_2\} \notin E$
- $\ldots$
- $p_y, q_y \in C - \{p_1, \ldots, p_{y-1}\}, p_y \neq q_y, \{p_y, q_y\} \notin E$

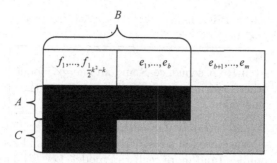

Fig. 5.22. Extracting clique C from bi-clique A, B

This tells us that there are at least $y+1$ different pairs of nodes in C without an edge connecting them. Thus, number b defined above can be estimated more strictly than in the former case as follows:

$$b \leq \frac{1}{2}(2k-a)(2k-a-1) - (y+1) < \frac{1}{2}(2k-a)(2k-a-1) - y .$$

From this we obtain a contradiction as follows (the first step uses $a = k-y > 0$, the last step uses $k \geq 4$, thus $-k-y+3 < 0$):

$$0 \leq |A|\,|B| - k^3 + \frac{3}{2}k^2$$

$$< (k-y)\left(\frac{1}{2}k^2 - k + \frac{1}{2}(k+y)(k+y-1) - y\right) - k^3 + \frac{3}{2}k^2$$

$$= \frac{1}{2}k^3 - k^2 - \frac{1}{2}yk^2 + yk + \frac{1}{2}\left(k^2 - y^2\right)(k+y-1) - (k-y)y - k^3 + \frac{3}{2}k^2$$

$$= \frac{1}{2}k^3 - k^2 - \frac{1}{2}yk^2 + yk + \frac{1}{2}k^3 + \frac{1}{2}yk^2 - \frac{1}{2}k^2 - \frac{1}{2}y^2k - \frac{1}{2}y^3 + \frac{1}{2}y^2$$

$$\quad - ky + y^2 - k^3 + \frac{3}{2}k^2$$

$$= -\frac{1}{2}y^2k - \frac{1}{2}y^3 + \frac{3}{2}y^2$$

$$= \frac{1}{2}y^2(-k - y + 3)$$

$$\leq 0 .$$

Thus the assumption that no clique of size k exists cannot be maintained. This proves the desired conclusion. □

5.5.6 Pseudoknot Prediction

Pseudoknot prediction is an *optimization problem*. Given an RNA sequence, we want to compute a possibly pseudoknotted structure with minimum free energy in regards to a certain energy model. Stated as a *decision problem*, pseudoknot prediction PK asks if a structure with free energy lower than a certain threshold exists.

> **PK**
> Given RNA sequence S and an energy model, is there a structure with free energy lower than some value E_{low}.

Let us fix a simple (and certainly unrealistic) *energy model*. This makes sense, as complexity results achieved in a restricted model will hold also in the corresponding more complex (and more realistic) model. Let R be an RNA structure over $S = [1 \ldots n]$, where pseudoknots are allowed. The *nearest neighbour pseudoknot model* defines energy $E(R)$ as the independent sum of base pair energies

$$E(R) = \sum_{(i,j) \in R} E(i, j, i+1, j-1) \, ,$$

where the energy of a base pair (i, j) depends on the types of the four bases $i, j, i+1, j-1$ and furthermore, if $(i+1, j') \in R$ (or $(i', j-1) \in R$), also possibly on base j' (or base i'). We will prove complexity of pseudoknot prediction in the nearest neighbour pseudoknot model and refer to this problem as PKNN.

> **PKNN**
> Given RNA sequence S and the nearest neighbour pseudoknot model, is there a structure with free energy lower than some value E_{low}.

For proving complexity of PKNN, a restriction of the well-known 3SAT problem is used. 3SAT remains NP-complete if it is required that each variable appears exactly two times positively and one or two times negatively. We refer to this restricted problem as 3SAT$_{\text{restricted}}$ and show its NP-completeness within the proof of the following theorem.

Theorem 5.43.
3SAT$_{\text{restricted}}$ *is polynomially reducible to PKNN, thus PKNN is NP-complete, too.*

Proof. We follow the proof presented in [52] with some minor modifications to enhance understanding.

First, we start with introducing the *energy function* for the underlying nearest neighbour pseudoknot model. Broadly speaking, negative (favourable) energy is assigned to complementary symbols forming base pairs if and only if there is no pseudoknotted interaction with a base pair involving a neighbouring symbol. This can be formalized as follows:

$$E(i, j, i+1, j-1) = \begin{cases} -1 & \text{if } i \text{ and } j \text{ are complementary symbols and for} \\ & k \notin \{i+1, \ldots, j-1\} \text{ there are no base pairs} \\ & (i+1, k), (j-1, k), (k, i+1), (k, j-1). \\ 0 & \text{otherwise.} \end{cases}$$

Figure 5.23 displays examples for different cases we can distinguish in the nearest neighbour pseudoknot model using the described energy function.

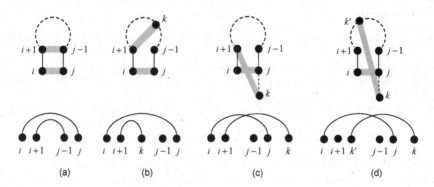

Fig. 5.23. Energy contribution of base pair (i,j) according to the underlying energy function: **(a)** stem with energy -1; **(b)** bulge with energy -1; **(c)** pseudoknot structure involving the *neighbouring* base $i+1$ penalized with energy 0; **(d)** pseudoknot structure not involving a *neighbouring* base, therefore no penalty and energy -1

We proceed with the reduction. First, we reduce 3SAT to 3SAT$_{\text{restricted}}$ where we demand that for every variable x there are exactly two occurrences of positive literal x and one or two occurrences of negative literal $\neg x$. Reduction to this restricted problem is done in an easy manner as follows. For example, assume that a variable x has four positive and three negative occurrences in φ. Replace these occurrences by fresh literals $y_1, y_2, y_3, y_4, \neg z_1, \neg z_2, \neg z_3$ and enforce that $y_1, y_2, y_3, y_4, z_1, z_2, z_3$ are representatives for the same x by the following implications, using a further auxiliary Boolean variable w.

$$y_1 \rightarrow y_2 \rightarrow y_3 \rightarrow y_4 \rightarrow z_1 \rightarrow z_2 \rightarrow z_3 \rightarrow w \rightarrow y_1$$

Now implications are transformed into disjunctions as follows.

$$\neg y_1 \vee \neg y_1 \vee y_2 \qquad \neg y_2 \vee \neg y_2 \vee y_3 \qquad \neg y_3 \vee \neg y_3 \vee y_4 \qquad \neg y_4 \vee \neg y_4 \vee z_1$$
$$\neg z_1 \vee z_2 \vee z_2 \qquad \neg z_2 \vee z_3 \vee z_3 \qquad \neg z_3 \vee w \vee w \qquad \neg w \vee \neg w \vee y_1$$

Note that z_1 occurs two times as a positive literal and only once as a negative literal, whereas all other variables occur two times as a positive literal and two times as a negative literal. Next we reduce 3SAT$_{\text{restricted}}$ to PKNN. Let an instance

$$\varphi = (L_{11} \vee L_{12} \vee L_{13}) \wedge \ldots \wedge (L_{n1} \vee L_{n2} \vee L_{n3})$$

of 3SAT$_{\text{restricted}}$ be given with m clauses and k variables $x_1, \ldots, x_k$. In the reduction we use some sort of "generalized RNA folding problem" where more complementary pairs of characters are available than only the standard complementary RNA bases forming pairs (A,U) and (C,G). This considerably simplifies reduction. Reduction using the 4-character alphabet A, C, G, U requires some additional efforts (see [52]). In this proof, complementary character pairs are always denoted by $x, \underline{x}$. The following complementary characters pairs are used:

- For every $i = 1, \ldots, m$ (m the number of clauses) two pairs $A_i, \underline{A}_i$ and $B_i, \underline{B}_i$
- For every $j = 1, \ldots, k$ (k the number of variables) one pair $V_j, \underline{V}_j$
- For every variable x_j two pairs $x_j, \underline{x}_j$ and $\neg x_j, \underline{\neg x}_j$

Now we encode i^{th} clause $(L_{i1} \vee L_{i2} \vee L_{i3})$ by the following *clause substring*:

$$C_i = A_i L_{i1} \underline{A}_i B_i L_{i2} A_i \underline{B}_i L_{i3} B_i.$$

Observe the alternation of characters $A_i, \underline{A}_i, A_i$ and correspondingly $B_i, \underline{B}_i, B_i$. Variable x_j is encoded by the following *variable substring*:

$$X_j = V_j \underline{x}_j x_j \underline{V}_j \neg x_j \underline{\neg x}_j V_j.$$

Finally we form *clause string* $C = C_1 \ldots C_m$, *variable string* $X = X_1 \ldots X_k$, and their concatenation $S = CX$.

Lemma 5.44. *If φ is satisfiable, then S has a fold with free energy $-3m - k$.*

Proof. Consider a truth-value assignment that satisfies φ. Within each clause choose an arbitrary literal L that evaluates to 'true' under the chosen assignment. If the chosen literal L is the first occurrence of L within clause string C, let it be paired with the second occurrence of complementary literal $\underline{L}$ within variable string X. If the chosen L is the second occurrence of L within clause string C, let it be paired with the first occurrence of complementary literal $\underline{L}$ within variable string X. This *"first-second rule"* guarantees that for the fixed literals within each clause their occurrence in string C and corresponding complementary occurrence in string X do not form unfavourable pseudoknots according to the energy function. Note that there might well be pairs that form a pseudoknot, but only favourable ones as the pairs forming a pseudoknot cannot be in contact.

So far, we have formed m pairs that contribute $-m$ to overall free energy. Within each string C_i there is the possibility to form two further favourable pairs between the pairs $A_i, \underline{A}_i$ and $B_i, \underline{B}_i$, respectively. These additional pairs contribute -2 to overall free energy for each clause substring. So far, free energy of the fold corresponding to a truth-value assignment that satisfies φ has value $-3m$. Finally, we introduce one further favourable pair within each variable substring X_j as follows. If literal x_j becomes 'true' by our truth-value assignment, a pair between symbol $\underline{V}_j$ and second occurrence of complementary symbol V_j in variable substring X_j is formed. If literal $\neg x_j$ becomes 'true' by our truth-value assignment, a pair between symbol $\underline{V}_j$ and first occurrence of complementary symbol V_j in variable substring X_j is formed. As a consequence, we prevent that favourable pairs are formed involving literals that become 'false' by our truth-value assignment. Note that each of these additional pairs contributes -1 to overall free energy. Summarizing, the constructed pairs lead to an overall free energy for the fold of $-3m - k$.   □

As an example, consider the 3SAT$_{\text{restricted}}$ formula

$$\varphi = (x \vee \neg y \vee \neg x) \wedge (z \vee \neg y \vee y) \wedge (\neg x \vee y \vee x) \wedge (\neg z \vee z \vee \neg z)$$

with four clauses and three variables. As requested, each literal appears at most twice. Corresponding clause and variable substrings and construction of the fold with energy $-3m - k$ for φ are shown in Fig. 5.24.

Lemma 5.45. *If S has a fold with free energy $-3m - k$, then φ is satisfiable.*

Proof. The fold constructed as described by the rules above is the only one that achieves a free energy of $3m - k$. This can be shown as follows. Whatever pairings we form, every clause substring C_i contributes at most three favourable pairs. Besides the three variants used in Lemma 5.44, other further pairings are possible (see Fig. 5.25). Whichever one we choose, it never results in more than three favourable pairs. Furthermore, in case that a clause substring exhibits three bonds, at least one of the links must lead out of the substring C_i. In order to achieve free energy value $-3m - k$ there must be at least one further favourable pair forming within each variable substring X_j. Looking at X_j we observe that the one remaining possible internal pairing $V_j, \underline{V}_j$ is favourable only in case that there are no links leading both to characters $\underline{x}_j$ as well as to $\underline{\neg x}_j$. This induces a truth-value assignment as follows: set x_j to 'true' if there is no link leading to $\underline{\neg x}_j$, otherwise set x_j to 'false'.

Thus we end with a fold exactly as constructed in Lemma 5.44. Since each clause substring C_j contains a link from some occurrence of literal L to some occurrence of $\underline{L}$ in the corresponding variable substring, we have defined a truth-value assignment that makes at least one literal within each clause 'true'. Consequently, φ is satisfiable.   □

Taking together the proofs for Lemma 5.44 and 5.45, we have shown that PKNN is NP-complete.   □

5.5.7 The Picture of Presented Reductions

... is shown in Fig. 5.26.

5.5.8 Further NP-Complete Bioinformatics Problems

We end with a short listing of a few NP-complete problems which are closely related to problems that we have treated in former chapters and give some bibliographic links. The reader who has got some familiarity with reductions and NP-completeness proofs after reading this chapter should have no problem to work through further such proofs in the literature.

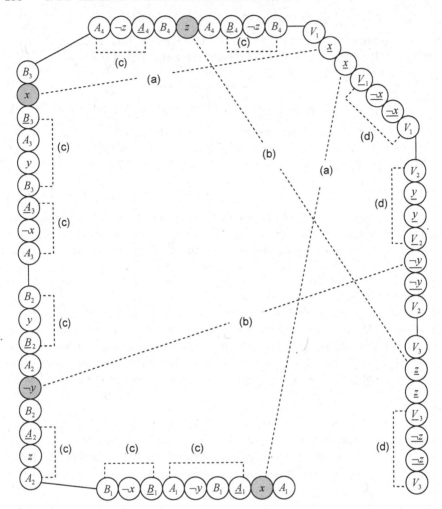

Fig. 5.24. Representative example fold illustrating proof of Lemma 5.44: literals x, $\neg y$ and z are set to 'true'; within each clause substring one 'true' literal is chosen (grey shaded); **(a)** "first-second rule" avoids unfavourable bonds; **(b)** pseudo-knots may arise, but always consisting of favourable bonds; **(c)** within each clause substring two additional favourable bonds are possible; **(d)** within each variable substring one additional favourable bond is possible.

- *Consensus (Steiner) String* (for any metric as distance function) [51, 69]
- *Undirected Genome Rearrangement* [15]
- *HP-model fold optimization* [11]

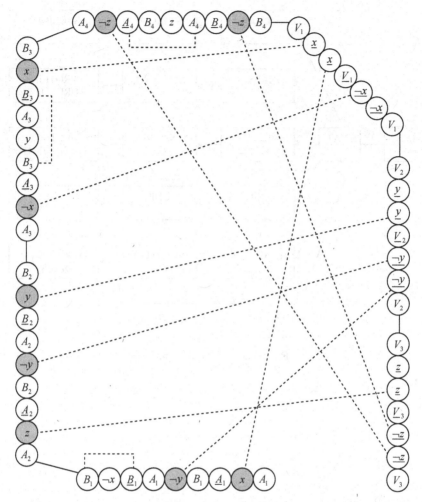

Fig. 5.25. Representative example fold illustrating proof of Lemma 5.45: every clause substring contributes at most three favourable bonds (some further situations besides the ones shown in Fig. 5.24 are exhibited); a clause substring cannot contain three favourable inner bonds; using bonds to complementary literals within some variable substring prevents formation of further favourable inner bonds within that variable substring

5.6 Bibliographic Remarks

To the personal opinion of the author, the best introductory textbook to the field of complexity theory, and in particular to the concept NP-completeness is Papadimitriou [61]. Of course, any treatment of NP-completeness must mention the absolutely classical textbook (nice to read and informative up to the present time) Garey & Johnson [29].

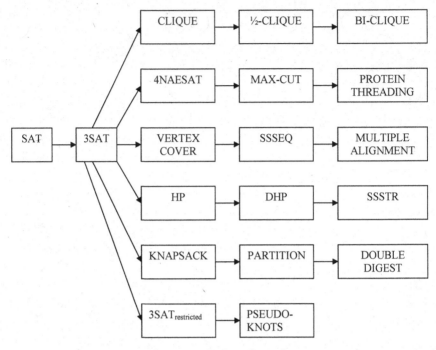

Fig. 5.26. Reduction paths

6

Approximation Algorithms

6.1 Basics of Approximation Algorithms

6.1.1 What is an Approximation Algorithm?

The standard form of algorithmic problems studied in this book so far was a minimization or maximization problem. In the following we concentrate on minimization, knowing that maximization can be treated completely similar. Given an admissible instance x, a solution y is to be found having minimum value $c(x, y)$ among all solutions for x. As we did in Chap. 5 we also assume here that for every instance $x \in I$ there exists at least one y with $(x, y) \in S$. An exact algorithm A returns for every instance x a solution $y = A(x)$ for x with $c(x, A(x)) = c_{\min}(x)$ with

$$c_{\min}(x) = \min_{\substack{\text{all solutions } y \\ \text{for instance } x}} c(x, y) . \tag{6.1}$$

For a fixed parameter $\alpha \geq 1$, an α-approximation algorithm B computing a solution $y = B(x)$ for every admissible instance x relaxes this requirement to

$$c(x, B(x)) \leq \alpha c_{\min}(x) . \tag{6.2}$$

The closer α to 1 is, the better is the approximation solution $B(x)$. In the extreme case of $\alpha = 1$ we even have an exact algorithm.

6.1.2 First Example: Metric Travelling Salesman

Consider an undirected weighted graph consisting of n completely intercon-nected nodes with a positive distance value $d_{uv} = d_{vu}$ assigned to each edge $\{u, v\}$. Assume that the distance values satisfy the triangle inequality $d_{uv} \leq d_{uw} + d_{wv}$. Then, a closed tour, that is, a cycle through the graph having costs not greater that two times the minimum possible distance sum can be efficiently constructed as follows:

- Compute a subset of edges that forms a tree, covers every node of the graph, and has minimum costs (sum of distances assigned to the edges of the tree) among all such trees. Trees with these properties are well-known under the name of *minimum spanning trees*. They can be easily constructed by a rather trivial greedy approach that extends a growing tree at every stage by a fresh edge with minimum edge costs. Figure 6.1 shows a graph and a minimum spanning tree (its edges shown by bold lines). Imagine this example to be a computer network with varying communication costs between computers; messages may pass forever through the network due to the presence of lots of cycles; a spanning tree guarantees that every message send by any computer arrives at each other computer without passing forever through the network; a minimum spanning tree does this at cheapest costs.
- Given a cheapest Hamiltonian cycle with cost c_{cycle}, the deletion of an arbitrary edge from the cycle leads to a (special sort of non-branching) spanning tree with costs less than c_{cycle}. Thus the minimum costs of any spanning tree $c_{spanning}$ is also less than c_{cycle}.
- Now make a pre-order traversal through the computed spanning tree that starts at some arbitrary node and ends at that node. This traversal uses every edge of the spanning tree exactly twice (one time forth, and one time back), thus its costs are at most two times $c_{spanning}$. Unfortunately, it is not a Hamiltonian cycle since nodes are visited in a multiple manner. (Question: How often are nodes visited?). A Hamiltonian cycle is easily obtained from the pre-order traversal by avoiding multiple node visits by using shortcuts to the actually next so far unvisited node. Triangle inequality guarantees that this does not lead to increased costs.
- Taken together, the Hamiltonian cycle that is computed this way has costs no greater that $2c_{spanning}$, that is, no greater than $2c_{cycle}$.

Exercises

6.1. Apply a greedy algorithm and show how the minimum spanning tree from Fig. 6.1 evolves.

6.2. Try to find out how a simple improvement of the construction above leads to a 3/2-approximation algorithm. Use a more clever way to make a tree traversal through the graph.

6.1.3 Lower and Upper Bounding

As the example above showed, the 2-approximation algorithm centers around the concept of a feasible, that is efficiently computable lower bound $c_{spanning}$ for the desired, but unfeasible minimum value c_{cycle}. Instead of the desired estimation $c(x, A(x)) \leq 2c_{cycle}(x)$ we even show $c(x, A(x)) \leq 2c_{spanning}(x)$. To define, in the general case, an appropriate feasible lower bound $c_{lower}(x)$

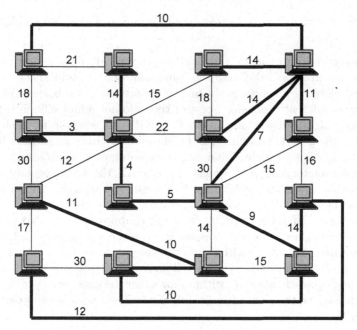

Fig. 6.1. Minimal costs spanning tree

for $c_{\min}(x)$ is a delicate task. First note that, for example for the case of a 2-approximation algorithm any lower bound must necessarily be located in the interval between $1/2\, c_{\min}(x)$ and $c_{\min}(x)$. Being not less than $1/2\, c_{\min}(x)$ is necessary since otherwise the intended estimation $c(x, A(x)) \leq 2c_{\mathrm{lower}}(x)$ would not be possible. Both extreme values $1/2\, c_{\min}(x)$ and $c_{\min}(x)$ cause difficulties. If $c_{\mathrm{lower}}(x)$ is too close to $c_{\min}(x)$, it might be as hard to compute as $c_{\min}(x)$ is. If $c_{\mathrm{lower}}(x)$ is too close to $1/2\, c_{\min}(x)$ the computed solutions are almost optimal, thus eventually not efficiently computable. Summarizing, relaxation of $c_{\min}$ to some lower bound c_{lower} as basis for an approximation algorithm must be strong enough to lead to an efficiently computable bound, but must not be so strong that computed solutions are too close to optimal solutions, thus eventually not efficiently computable.

6.1.4 What are Approximation Algorithms good for?

Having an approximation algorithm that delivers solutions at most 1% worse than optimum solutions, that is a 1.01-approximation algorithm, usually is welcome as an approximation algorithm. For the travelling salesman problem we have presented a 3-approximation algorithm, and in Sect. 6.2 we will present a 4-approximation algorithm for the shortest common superstring problem. Can you really imagine that a bus driver would be happy about a computed tour through town that is at most 3 times as long as a shortest possible tour? We try to give some reasons why even a 4-approximation algorithm might be welcome.

(1) Concerning approximation algorithms, the world of NP-complete problems that presents itself from the standpoint of exact solutions as a world of completely equivalent problems, splits into various subclasses of qualitatively different problems. Some of them do not admit efficient approximation algorithms at all (unless P = NP). An example for such problems is the general travelling salesman problem. Other problems admit α-approximation algorithms, but only with parameters up to some smallest such parameter $\alpha_{min} > 1$. An example is 3SAT, the well-known satisfiability problem for Boolean formulas in clausal form with exactly three literals per clause. The provably best α-approximation algorithm has $\alpha = 7/8$ (unless P = NP). Other problems admit α-approximation algorithms for every fixed parameter $\alpha > 1$, some even in the strong manner that α is a parameter of the algorithm and its running time scales with $1/\alpha$. An example for such problems is the well-known knapsack problem. Thus having an α-approximation algorithm shows that the considered problem does not belong to the class of most difficult problem with no approximation algorithm at all.

(2) Having an α-approximation algorithm often opens way to find also a β-approximation algorithm with improved parameter $\beta < \alpha$. As an example, the 4-approximation algorithm that will be presented in Sect. 6.2 can easily be improved to a 3-approximation algorithm, and with some effort even to a 2.75-approximation algorithm.

(3) For the 4-approximation algorithm that will be presented in Sect. 6.2, estimation of the costs of computed solutions by four times the minimal costs is only an upper bound. Computed solutions may often be much closer to minimum cost solutions. Having computed a solution $A(x)$ its costs might be, in extreme cases, close to $c_{min}(x)$ (that might seem to be preferable), or close to $4c_{min}(x)$ (that might seem not to be preferable). Unfortunately, how good a computed solution $A(x)$ really is cannot be detected, since we have no knowledge on $c_{min}(x)$. There are reasons that even computed solutions of bad quality may be of some value. To explain why imagine that we had two solutions $A(x)$ and $B(x)$ for the same instance x. Now consider the case that $A(x)$ is rather bad having worst costs $4c_{min}(x)$. Though we cannot detect this, this might help to detect that $B(x)$ is a much better solution, for example if we observe that $c(x, B(x)) < 1/3\, c(x, A(x))$ holds. Together with the guaranteed estimation $c(x, A(x)) \leq 4c_{min}(x)$ we then obtain $c(x, B(x)) < 4/3\, c_{min}(x)$. This is considerably closer to optimum - and has been detected.

6.2 Shortest Common Superstring Problem

6.2.1 Feasible Lower Bound

The SHORTEST COMMON SUPERSTRING problem was introduced and illustrated in Sect. 2.3. We refer to the concepts and notions introduced there. Having a shortest superstring S for a substring-free string set F, this defines a unique ordering $F_1, F_2, \ldots, F_m$ of strings from F according to their start position within S. Appending first string F_1 to the end of this sequence, we obtain a situation as shown in Fig. 6.2 for the case of four fragments. In general, S can be decomposed as follows, with length l being the sum of m prefix lengths and a single overlap length:

$$S = \mathrm{prefix}(F_1, F_2) \ldots \mathrm{prefix}(F_{m-1}, F_m)\mathrm{prefix}(F_m, F_1)\mathrm{overlap}(F_m, F_1)$$
$$l = p(F_1, F_2) + \ldots + p(F_{m-1}, F_m) + p(F_m, F_1) + o(F_m, F_1) \,. \tag{6.3}$$

Conversely, any cycle $F_1, F_2, \ldots, F_m, F_1$ induces a string S as defined above to which we refer as the string defined by cycle $F_1, F_2, \ldots, F_m, F_1$. Note that costs c of cycle $F_1, F_2, \ldots, F_m, F_1$ in prefix graph are given by the slightly smaller value than length l

$$c = p(F_1, F_2) + \ldots + p(F_{m-1}, F_m) + p(F_m, F_1) \,. \tag{6.4}$$

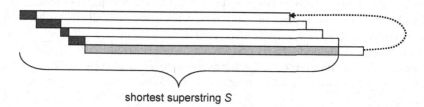

shortest superstring S

Fig. 6.2. Cycle of prefixes

This immediately gives us the following lower bound.

Lemma 6.1. *Lower Bound*
For a substring-free string set F let $\pi(F)$ be the minimum costs of any cycle in prefix graph P_F, and $\sigma(F)$ be the minimum length of any common superstring for set F. Then $\pi(F) \le \sigma(F)$.

Unfortunately, lower bound $\pi(F)$ is as difficult to compute as $\sigma(F)$ because its computation corresponds to the well-known NP-complete travelling salesman problem. An explanation for this might be, that $\pi(F)$ is a too strong lower bound, that is, too close to $\sigma(F)$. Fortunately, there is a weaker well-known lower bound. This is based on a relaxation of the travelling salesman problem that admits covering nodes by several disjoint local cycles instead by a single one. A finite set of disjoint cycles that contains every node of a weighted graph is called a *cycle cover* of the graph. Costs of a cycle cover is defined to be the sum of costs of each cycle in the cycle cover. For a substring-free string set F denote by $\gamma(F)$ the minimum costs of any cycle cover in prefix graph P_F.

Lemma 6.2. *Feasible Lower Bound*
For a substring-free string set F the following holds: $\gamma(F) \leq \pi(F) \leq \sigma(F)$.

Proof. $\gamma(F) \leq \pi(F)$ follows from the fact that a single cycles is a special case of a cycle cover. □

6.2.2 Cycle Covers Related to Perfect Matchings Problem

Here we treat the problem of computing a cycle cover with minimum costs in prefix graph. As for any cycle $F_1, F_2, \ldots, F_m, F_1$ the sum of prefix lengths $p(F_1, F_2) + \ldots + p(F_{m-1}, F_m) + p(F_m, F_1)$ and the sum of overlap lengths $o(F_1, F_2) + \ldots + o(F_{m-1}, F_m) + o(F_m, F_1)$ sum up to the sum of lengths of strings $F_1, \ldots, F_m$ (a constant value independent of the considered cycle), we may equivalently consider the problem of computing a *cycle cover with maximum costs in overlap graph*. Using overlap graph instead of prefix graph will prove to be more convenient in the following.

Next we transform the cycle cover problem for overlap graph into a perfect matching problem in a *bipartite version* of overlap graph. The latter is defined as follows. Create for every node F_i in overlap graph a copy node called G_i. Thus the new graph consists of two parts, a left part with all the nodes from the original graph, and a right part that is a copy of the left part. Every directed edge from node F_i to node F_j in overlap graph with weight o_{ij} is simulated by an undirected edge between node F_i and copy node G_j with same weight o_{ij}. Now consider an arbitrary local cycle with costs c in overlap graph.

$$F_1 \rightarrow F_2 \rightarrow F_3 \rightarrow \ldots \rightarrow F_{m-1} \rightarrow F_m \rightarrow F_1$$

Its directed edges correspond to undirected edges of the bipartite version as follows:

$$
\begin{array}{ccccccc}
F_1 & F_2 & F_3 & \ldots & F_{m-1} & F_m \\
| & | & | & & | & | \\
G_2 & G_3 & G_4 & \ldots & G_m & G_1
\end{array}
$$

Such a one-to-one relation between node sets $\{F_1, \ldots, F_m\}$ and $\{G_1, \ldots, G_m\}$ with an undirected edge between any two related nodes is called a *matching*. The costs of a matching are defined as the summed weights of its undirected edges. We observe that the costs of the constructed matching coincide with the costs c of the considered local cycle. Conversely, having a matching with costs c between node sets $\{F_1, \ldots, F_m\}$ and $\{G_1, \ldots, G_m\}$ we may always arrange matches pairs in an ordering as above, thus we obtain a local cycle with costs c through node set $\{F_1, \ldots, F_m\}$. Now let us consider an arbitrary cycle cover with costs c in overlap graph. Its cycles lead to a collection of (local) matchings that together form a matching with costs c, called a *perfect matching* ("perfect" since all nodes participate in the matching).

6.2.3 Computing Maximum Cost Perfect Matchings

Reduction of the minimum costs cycle cover problem for prefix graphs to the maximum costs perfect matching problem in bipartite versions of overlap graphs has the advantage that there are well-known efficient (quadratic) algorithms solving perfect matching problems in arbitrary bipartite graphs. We must not rely on such algorithms since situation here is even better. For an overlap graph we show that its bipartite version fulfils a certain property called *Monge property*. This property will be the basis for a most simple greedy algorithm for the construction of a maximum costs perfect matching.

Lemma 6.3. *Monge Property*
Consider in the bipartite overlap graph four different nodes, two F_a and F_b on the left part of the graph, and two G_c and G_d on the right part of the graph. Assume that $o(F_a, G_c)$ is maximum among $o(F_a, G_c)$, $o(F_a, G_d)$, $o(F_b, G_c)$ and $o(F_b, G_d)$. Then the following holds:

$$o(F_a, G_c) + o(F_b, G_d) \geq o(F_a, G_d) + o(F_b, G_c)$$

Proof. Since $o(F_a, G_c) \geq o(F_a, G_d)$ and $o(F_a, G_c) \geq o(F_b, G_c)$ hold, the diagram in Fig. 6.3 correctly indicates relative locations of overlaps (overlap between F_a and G_c is shown as black shaded area, the shorter overlaps between F_b and G_c and between F_a and G_d are shown as grey shaded area). Now the desired inequality $o(F_a, G_c) + o(F_b, G_d) \geq o(F_a, G_d) + o(F_b, G_c)$ can be easily extracted from the diagram (already the guaranteed minimum overlaps between F_b and G_d suffices to give the inequality). $\square$

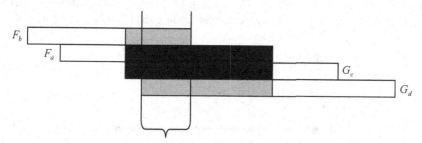

Fig. 6.3. F_b and G_d overlap at least that much.

As a consequence we next show that the following greedy algorithm indeed computes a maximal matching. The algorithm uses a growing set of already matched node pairs, and decreasing sets of still unmatched nodes for both parts of the bipartite overlap graph.

$M = \emptyset$
$V =$ all nodes in left part of the bipartite graph
$W =$ all nodes in right part of the bipartite graph
while $V \neq \emptyset$ **do**
    choose F in V and G in W having maximal value $o(F, G)$
    $V = V - \{F\}$
    $W = W - \{G\}$
    $M = M \cup \{\{F, G\}\}$
end while

Lemma 6.4.
For an arbitrary bipartite graph with $2n$ nodes that fulfils the Monge property the greedy algorithm above always returns a maximal matching in $O(n^2 \log n)$ steps.

Proof. We compare the matching M that is computed by the algorithm and an arbitrary maximal weight matching M^*. Following the order in which node pairs are placed into M by the greedy algorithm, we show how to successively reorganize the maximum matching M^* into a maximum matching M^{**} in such a way that optimality is always preserved and finally the same node pairs appear in M and in M^{**}. Thus, at the end of this reorganization process, M is shown to coincide with optimal matching M^{**}, that is M itself is shown to be optimal.

Consider some pair $\{F_a, G_c\}$ at the moment it was placed into M by the greedy algorithm. Assume that M and M^{**} coincide on all pairs that were placed into M before this moment. Also assume that the actually chosen pair $\{F_a, G_c\}$ is not present in M^{**}. Thus M^{**} must contain different pairs $\{F_a, G_d\}$ and $\{F_b, G_c\}$. Now also consider node pair $\{F_b, G_d\}$ (see Fig. 6.4).

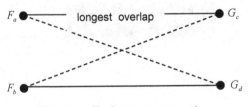

Fig. 6.4. Exchange construction

Since M and M^{**} coincide on all pairs placed into M before the moment of insertion of $\{F_a, G_c\}$, all of these four pairings were available for the greedy algorithm at the moment of insertion of $\{F_a, G_c\}$. As $\{F_a, G_c\}$ had maximum weight among these four pairs by the way the algorithm worked, we know that $o(F_a, G_c) + o(F_b, G_d) \geq o(F_a, G_d) + o(F_b, G_c)$ holds by the Monge property. Replacing pairs $\{F_a, G_d\}$ and $\{F_b, G_c\}$ in M^{**} by pairs $\{F_a, G_c\}$ and $\{F_b, G_d\}$, overall weight of M^{**} does not decrease, and of course cannot increase by optimality of M^{**}. Thus optimality of M^{**} is preserved. In addition, M and M^{**} now share the node pair $\{F_a, G_c\}$ actually placed into M. Execution time $O(n^2 \log n)$ for the algorithm results by first sorting the n^2 weights of edges in time $O(n^2 \log n^2) = O(n^2 \log n)$. Then the process of building matching M requires a linear walk through at most $O(n^2)$ edges.   □

6.2.4 Cycle Cover Versus Cyclic Covering by Strings

The concatenation of prefix strings introduced above (in Fig. 6.2 shown for the case of a cycle F_1, F_2, F_3, F_4, F_1 of length 4 by black coloured areas, whereas the final overlap string is indicated by a grey coloured area) plays an important role in the following discussion. This string is called the *prefix string* P defined by the considered local cycle.

$$P = \text{prefix}(F_1, F_2)\text{prefix}(F_2, F_3) \dots \text{prefix}(F_{m-1}, F_m)\text{prefix}(F_m, F_1) \quad (6.5)$$

Lemma 6.5.
Prefix string P of a local cycle $F_1, \dots, F_m, F_1$ is a non-empty string. Strings $F_1, \dots, F_m$ occur as substrings of the m-fold concatenation P^m of P, for a sufficiently great number m. It is said that string P cyclically covers a string set in case that all strings in the covered string set occurs as substrings of the m-fold concatenation P^m of P, for a sufficiently great number m.

Proof. Take a look at the situation shown in Fig. 6.5. We observe that prefix string P, i.e. the concatenation of the four black coloured areas forms a non-empty string (remember that overlaps were defined as proper overlaps, also for the case of an overlap of a string with itself). Also P occurs as a prefix of F_1. Thus, P occurs a second time in the diagram as prefix of the second occurrence of F_1 shown as grey coloured area. Switching to the first occurrence of F_1 entails that concatenation PP occurs as a prefix F_1. Switching back to the second occurrence of F_1 we also find PP as prefix of the grey coloured area. When applying these back-and-forth switches sufficiently often the conclusion of the lemma follows.   □

Having shown that for a local cycle $F_1, \dots, F_m, F_1$ its prefix string P cyclically covers every strings within string set $\{F_1, \dots, F_m\}$, we conversely show that any string Q that cyclically covers a string set leads to an ordering $F_1, \dots, F_m$ of the strings in the covered string set with prefix string P that is no longer than Q. This can be seen as follows. Assume that strings within the considered string set occur as substrings of Q^m, for some number m. Arrange strings from the considered string set according to their start position within the left-most

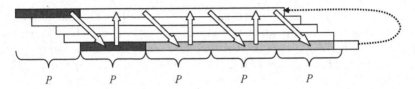

Fig. 6.5. Prefix string P cyclically covers strings from a local cycle

copy Q. We obtain a permutation $F_1, \ldots, F_m$ of the strings. Denote by p the start position of string F_1 in the first copy of Q. Now note that the second use of string F_1 as end string in cycle $F_1, \ldots, F_m, F_1$ starts before or at the same position p in the second copy of Q. The prefix string P defined by this ordering is thus located between position p in the first copy of Q and some position q left of or equal to position p in the second copy of Q. Thus, P is no longer than Q is.

This observation opens way for a different interpretation of a minimum cycle cover for a string set. Given a minimum cycle cover of size σ for string set F that consists of k local cycles with prefixes $P_1, \ldots, P_k$ of lengths $p_1, \ldots, p_k$ fulfilling $\sigma = p_1 + \ldots + p_k$, we obtain the string set $C = \{P_1, \ldots, P_k\}$ of overall size $\sigma = p_1 + \ldots + p_k$ such that every string in F is cyclically covered by at least one of the strings from C. We briefly say that C *cyclically covers* F. Thus the minimum overall size of a string set that cyclically covers F is less or equal to σ. Conversely, having a string set C of minimum overall size σ that cyclically covers F, we may construct from this a finite set of local cycles that forms a cycle cover for F with prefixes $P_1, \ldots, P_k$ of lengths $p_1, \ldots, p_k$ fulfilling $\sigma \leq p_1 + \ldots + p_k$. Thus the minimum sum of lengths in a cycle cover for F is less or equal to $p_1 + \ldots + p_k$.

Summary

For a substring-free string set F, there is a one-to-one correspondence between minimum size cycle covers in the prefix graph of F, and minimum size strings sets that cyclically cover all strings of F. Thus we may freely use the model that is best suited in every of the forth-coming constructions.

6.2.5 Some Combinatorial Statements on Cyclic Covering

We say that string S has *period* p if for all i such that $1 \leq i \leq |S|$ and $1 \leq i + p \leq |S|$, character $S(i)$ coincides with character $S(i+p)$. Let $\gcd(p, q)$ denote the *greatest common divisor* of two positive integers p and q.

Lemma 6.6. *Greatest Common Divisor Property*

If string S of length at least $p+q$ has period p as well as period q, then S also has $\gcd(p, q)$ as period.

Proof. If $p = q$ then $\gcd(p, q) = p$ and the assertion of the lemma is trivial. Now assume $q < p$. Since Euklid's algorithm computes $\gcd(p, q)$ by successively subtracting the smaller number from the greater one, it suffices to show that S also has period $p - q$. Consider i such that $1 \le i \le |S|$ and $1 \le i + p - q \le |S|$. If $i \le q$ then $1 \le i + p \le q + p \le |S|$ holds. As S has period p and also period q we obtain $S(i) = S(i + p) = S(i + p - q)$. If $i > q$ then $1 \le i - q \le |S|$ holds. As S has period q and also period p we obtain $S(i - q) = S(i) = S(i - q + p)$. In both cases we have shown that $S(i) = S(i + p - q)$ holds. $\qquad\square$

Lemma 6.7. *Cyclic Shift Property*
Let string S be decomposed as $S = AB$. Then string $T = BA$ is called a cyclic shift *of S. Strings S and T cyclically cover the same strings.*

Proof. Note that $(AB)^m$ is a substring of $(BA)^{m+1}$, and $(BA)^m$ is a substring of $(AB)^{m+1}$. $\qquad\square$

Lemma 6.8. *Overlap Property*
Consider a minimum size cyclic cover C for a substring-free string set F. Let P and P' be different strings in C. Let S be cyclically covered by P and S' be cyclically covered by P'. Then the following estimation holds:

$$o(S, S') < |P| + |P'|$$

Proof. Assume that $o(S, S') \ge |P| + |P'|$ holds. We treat the case that $|P'| \le |P|$ holds. Since overlap(S, S') is a substring of both S and S', and since S and S' are cyclically covered by P and P', respectively, we know that overlap(S, S') has period $|P|$ as well as period $|P'|$.

We show that $|P| \ne |P'|$ holds. Assume that $|P| = |P'|$ holds. Since $o(S, S') \ge |P| + |P'|$ holds, we may assume that some cyclic shift Q of P starts at the first character of overlap(S, S') and is completely contained within overlap(S, S'). Similarly, we may assume that some cyclic shift Q' of P' starts at the first character of overlap(S, S') and is completely contained within overlap(S, S'). Thus $Q = Q'$. Replacing P and P' in cyclic cover C by the single string Q would give us a cyclic cover of string set F of lower size. This contradicts minimality of C.

Thus we have shown that $|P'| < |P|$ holds. Now we apply the *Greatest Common Divisor Lemma* and conclude that overlap(S, S') also has the greatest common divisor r of $|P|$ and $|P'|$ as period. Being a substring of overlap(S, S') we conclude that also P has period r. Consider the prefix R of P with length r. R is strictly shorter than P, and every string that is cyclically covered by P is also cyclically covered by R. Replacing string P by string R in the cyclic cover C leads to a new cyclic cover of smaller size. Again, this contradicts minimality of the given cyclic cover. $\qquad\square$

6.2.6 4-Approximation Algorithm

Let be given a substring-free set of strings F.

- Compute its overlap graph (using suffix trees as presented in Sect. 4.3 to achieve maximum efficiency).
- Compute a maximum weight perfect matching in overlap graph by the greedy algorithm presented above.
- Transform this maximum weight perfect matching into a minimum cost cycle cover in prefix graph. Let there be k local cycles $C_i = A_i, \ldots, B_i, A_i$ with initial string A_i and final string B_i (for $i = 1, \ldots, k$).
- Let P_i be the prefix string for cycle C_i, and $O_i = \text{overlap}(B_i, A_i)$. Return as common superstring S the concatenation of local common superstrings $P_i O_i$ (in arbitrary order).

As $P_i O_i$ is a local superstring for the strings in C_i, it follows that S is a superstring for all strings in F. Denote by c the costs of the computed minimum cost cycle cover.

$$c = |P_1| + |P_1| + \ldots + |P_m| \tag{6.6}$$

Let S_{opt} be a shortest superstring for F. We know that $c \leq |S_{\text{opt}}|$ holds. Furthermore we obtain:

$$
\begin{aligned}
|S| &= |P_1| + o(B_1, A_1) + \ldots + |P_m| + o(B_m, A_m) \\
&\leq c + |B_1| + \ldots + |B_m|
\end{aligned}
\tag{6.7}
$$

Now we also need a shortest superstring T_{opt} for the set of end strings $\{B_1, \ldots, B_m\}$ of the computed local cycles. Since S_{opt} is also a superstring for $\{B_1, \ldots, B_m\}$, we conclude that

$$|T_{\text{opt}}| \leq |S_{\text{opt}}| . \tag{6.8}$$

Arrange strings in $\{B_1, \ldots, B_m\}$ according to their start position in T_{opt} obtaining a permutation $B_{\pi(1)}, B_{\pi(2)}, \ldots, B_{\pi(m)}$. We thus know that:

$$
\begin{aligned}
|T_{\text{opt}}| &= p(B_{\pi(1)}, B_{\pi(2)}) + \ldots + p(B_{\pi(m-1)}, B_{\pi(m)}) + |B_{\pi(m)}| \\
&= |B_{\pi(1)}| - o(B_{\pi(1)}, B_{\pi(2)}) + \ldots + |B_{\pi(m-1)}| \\
&\quad - o(B_{\pi(m-1)}, B_{\pi(m)}) + |B_{\pi(m)}| .
\end{aligned}
\tag{6.9}
$$

From the *Overlap Lemma* we know that:

$$o(B_{\pi(i)}, B_{\pi(i+1)}) \leq |P_{\pi(i)}| + |P_{\pi(i+1)}| . \tag{6.10}$$

Taking all these observations together we obtain:

$$|S| \le c + |B_1| + \ldots + |B_m|$$
$$= c + |B_{\pi(1)}| + \ldots + |B_{\pi(m-1)}| + |B_{\pi(m)}|$$
$$= c + |T_{\text{opt}}| + o(B_{\pi(1)}, B_{\pi(2)}) + \ldots + o(B_{\pi(m-1)}, B_{\pi(m)})$$
$$\le c + |T_{\text{opt}}| + |P_{\pi(1)}| + |P_{\pi(2)}| + |P_{\pi(2)}| + |P_{\pi(3)}| \qquad (6.11)$$
$$+ \ldots + |P_{\pi(m-1)}| + |P_{\pi(m)}|$$
$$\le c + |T_{\text{opt}}| + c + c$$
$$\le 3c + |S_{\text{opt}}|$$
$$\le 4|S_{\text{opt}}| \ .$$

6.2.7 Putting Things Together

The central idea in the 4-approximation algorithm is to let local cycles (corresponding to growing local matchings) grow according to a greedy strategy in an óverlap graph. Whenever it happens that in such a growing local cycle $F_1, \ldots, F_m$ first string F_1 has greater overlap with last string F_m than any other not yet used string has with F_m, the local cycle must be closed. Instead of managing growing local cycles we may alternatively maintain the growing superstring S finally resulting from such cycles by concatenation. Note that the overlap between F_m and another string T is the same as the overlap between S and T, and the overlap between F_m and F_1 is also the same as the overlap between S and S itself. This leads to the following final formulation of the approximation algorithm, given a substring-free string set F. It uses variable $Open$ to maintain concatenated strings corresponding to still growing local cycles, and variable $Closed$ to store concatenated strings corresponding to closed local cycles.

```
Open = F
Closed = ∅
while Open ≠ ∅ do
    choose S and T in Open with maximal value o(S, T)
    if S ≠ T then
        merge S and T to obtain string merge(S, T)
        Open = (Open − {S, T}) ∪ {merge(S, T)}
    else
        Open = Open − {S}
        Closed = Closed ∪ {S}
    end if
end while
return arbitrary concatenation of all strings in Closed
```

Compare this pseudocode with the corresponding code for the originally discussed simple greedy algorithm for the shortest common superstring problem. It is astonishing to see that both codes differ only in a minor part. Nevertheless, the difference makes it possible to formally prove 4-approximation quality of the algorithm discussed here, whereas the exact status of the greedy algorithm is still unknown.

6.3 Steiner String

6.3.1 Problem Restated

The Steiner string problem is the task of finding a string S for given strings $S_1, \ldots, S_k$ which minimizes the pairwise alignment distance sum (using distance minimization instead of score maximization)

$$\sum_{i=1}^{k} d_{\text{opt}}(S_i, S) . \tag{6.12}$$

As was noted in Chap. 5, it is an NP-hard problem.

6.3.2 Feasible Lower Bound

Lemma 6.9.
Let S be a Steiner string for $S_1, \ldots, S_k$. Then there exists a string S_{next} among $S_1, \ldots, S_k$ such that the following estimation holds:

$$\sum_{i=1}^{k} d_{\text{opt}}(S_i, S_{\text{next}}) \leq \frac{2(k-1)}{k} \sum_{i=1}^{k} d_{\text{opt}}(S_i, S) .$$

This shows that we obtain a feasible lower bound for the Steiner sum 6.12 by taking a minimum:

$$\min_{a=1,\ldots,k} \frac{k}{2(k-1)} \sum_{i=1}^{k} d_{\text{opt}}(S_i, S_a) \leq \sum_{i=1}^{k} d_{\text{opt}}(S_i, S) .$$

Proof. For arbitrary index a, we estimate:

$$\sum_{i=1}^{k} d_{\text{opt}}(S_i, S_a) = \sum_{i=1, i \neq a}^{k} d_{\text{opt}}(S_i, S_a)$$

$$\leq \sum_{i=1, i \neq a}^{k} (d_{\text{opt}}(S_i, S) + d_{\text{opt}}(S, S_a))$$

$$= \sum_{i=1}^{k} d_{\text{opt}}(S_i, S) - d_{\text{opt}}(S_a, S) + (k-1)d_{\text{opt}}(S, S_a)$$

$$= \sum_{i=1}^{k} d_{\text{opt}}(S_i, S) - (k-2)d_{\text{opt}}(S, S_a) .$$

Now consider string S_{next} among $S_1, \ldots, S_k$ next to Steiner string S:

$$d_{\text{opt}}(S_{\text{next}}, S) = \min_{i=1,\ldots,k} d_{\text{opt}}(S_i, S) .$$

Note that S_{next} exits, whereas we cannot feasibly compute its distance to S. This will thus not be the lower bound we are looking for. By choice of S_{next} we know that

$$\sum_{i=1}^{k} d_{\text{opt}}(S, S_i) \geq k d_{\text{opt}}(S, S_{\text{next}}) .$$

Former estimation for $a = \text{next}$ and the latter estimation together yield:

$$\frac{\sum_{i=1}^{k} d_{\text{opt}}(S_i, S_{\text{next}})}{\sum_{i=1}^{k} d_{\text{opt}}(S_i, S)} \leq \frac{\sum_{i=1}^{k} d_{\text{opt}}(S_i, S) - (k-2) d_{\text{opt}}(S, S_{\text{next}})}{\sum_{i=1}^{k} d_{\text{opt}}(S_i, S)}$$

$$= 1 + \frac{(k-2) d_{\text{opt}}(S, S_{\text{next}})}{\sum_{i=1}^{k} d_{\text{opt}}(S_i, S)}$$

$$\leq 1 + \frac{(k-2) d_{\text{opt}}(S, S_{\text{next}})}{k d_{\text{opt}}(S_{\text{next}}, S)}$$

$$= 1 + \frac{k-2}{k} = \frac{2(k-1)}{k} .$$

Thus we have shown:

$$\sum_{i=1}^{k} d_{\text{opt}}(S_i, S_{\text{next}}) \leq \frac{2(k-1)}{k} \sum_{i=1}^{k} d_{\text{opt}}(S_i, S) .$$

$\square$

6.3.3 2-Approximation Algorithm

A *center string* of $S_1, \ldots, S_k$ is a string S_{center} taken from $S_1, \ldots, S_k$ which minimizes summed alignment distances to all other strings.

$$\sum_{i=1}^{k} d_{\text{opt}}(S_{\text{center}}, S_i) = \min_{a=1,\ldots,n} \sum_{i=1}^{k} d_{\text{opt}}(S_a, S_i)$$

Lemma 6.10.
The algorithm returning a center string S_{center} for $S_1, \ldots, S_k$ is a 2-approximation algorithm.

Proof.

$$\sum_{i=1}^{k} d_{\mathrm{opt}}(S_i, S_{\mathrm{center}}) \leq \min_{a=1,\dots,k} \sum_{i=1}^{k} d_{\mathrm{opt}}(S_i, S_a)$$

$$\leq \frac{2(k-1)}{k} \sum_{i=1}^{k} d_{\mathrm{opt}}(S_i, S)$$

$$\leq 2 \sum_{i=1}^{k} d_{\mathrm{opt}}(S_i, S)$$

$\square$

6.4 Phylogenetic Alignment

The construction presented above can be readily generalized to the phylogenetic alignment problem. Let a rooted tree be given with strings $S_1, \dots, S_k$ distributed to its leafs. We call an assignment of strings to its non-leaf nodes a *lifted phylogenetic alignment* if at every non-leaf node u some string attached to one of its sons is attached to u. This can also be described as follows: for every node u with attached string S_u there is a path from node u down to some leaf f such that the same string S_u is attached to all nodes along this path (and thus S_u coincides with the string S_f attached to the arrived leaf f).

Theorem 6.11. *Given a rooted tree with strings $S_1, \dots, S_k$ distributed to its leaves, there is a lifted phylogenetic alignment whose summed distance value is at most 2 times the minimum possible summed distance value of an arbitrary phylogenetic alignment.*

Proof. Let T_u be the string attached to non-leaf node u in a minimum distance phylogenetic alignment. In a bottom-up traversal through the tree we lift a certain string $S_u \in \{S_1, \dots, S_k\}$ to each non-leaf node u in such a way that the resulting attachment maximally resembles the optimal one. More precisely, working at non-leaf node u with successor nodes $u(1), \dots, u(k)$ which already got their lifted attachments $S_{u(1)}, \dots, S_{u(k)}$, we replace T_u with the string $S_{\mathrm{next}(u)}$ among $S_{u(1)}, \dots, S_{u(k)}$ that has least difference $d_{\mathrm{opt}}(S_{\mathrm{next}(u)}, T_u)$. Remember that such a choice of string $S_{\mathrm{next}(u)}$ also appeared in the proof for the Steiner approximation algorithm having approximation factor 2. Also note that here we only prove the existence of a lifted phylogenetic alignment of summed distance at most 2 times the minimum summed distance, the computation of such a lifted phylogenetic alignment was already performed in Chap. 3 using dynamic programming. Now we estimate the summed distance values within the lifted tree. For each leaf string S_a consider the path from S_a towards the root whose nodes are all labelled with the string S_a up to the last node v that is labelled with S_a. Consider also the immediate predecessor node u of v labelled with a string $S_{\mathrm{next}(u)}$ that is thus different from S_a. Note that the lifted phylogenetic tree is partitioned into these disjoint paths. Of course, only term $d_{\text{opt}}(S_{\text{next}(a)}, S_a)$ from each such path contributes to the summed distance value (all other distances have value 0). Thus situation is as shown in Fig. 6.6.

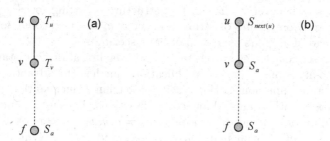

Fig. 6.6. (a) Minimum distance alignment; **(b)** one of the paths of the lifted alignment

The following estimations hold by choice of $T_{\text{next}(u)}$ and the triangle inequality:

$$d_{\text{opt}}(S_{\text{next}(u)}, T_u) \leq d_{\text{opt}}(S_a, T_u)$$
$$d_{\text{opt}}(S_{\text{next}(u)}, S_a) \leq d_{\text{opt}}(S_{\text{next}(u)}, T_u) + d_{\text{opt}}(T_u, S_a) .$$

Both together yield:

$$d_{\text{opt}}(S_{\text{next}(u)}, S_a) \leq 2 d_{\text{opt}}(T_u, S_a) .$$

Now we apply a cascade of triangle inequalities in the T-tree and obtain:

$$d_{\text{opt}}(T_u, S_a) \leq d_{\text{opt}}(T_u, T_v) + d_{\text{opt}}(T_v, S_a) \leq d_{\text{opt}}(\Pi_{u \to leaf})$$

where $\Pi_{u \to f}$ denotes the path from node u to leaf f and $d_{\text{opt}}(\Pi_{u \to f})$ the summed alignment distances along this path in the tree labelled with string T_u.

Taking all together we have shown:

$$d_{\text{opt}}(S_{\text{next}(u)}, S_a) \leq 2 d_{\text{opt}}(\Pi_{u \to f}) .$$

As we considered only pairs u, v with $S_u \neq S_v$, the occurring paths $\Pi_{u \to f}$ do not overlap. Summing over set X of all node pairs u, v in S with an edge from u to v and $S_u \neq S_v$ thus reduces to twice summing over set Y of all node pairs a, b in T with an edge from a to b:

$$\sum_{(u,v) \in X} d_{\text{opt}}(S_{\text{next}(u)}, S_v) \leq 2 \sum_{(a,b) \in Y} d_{\text{opt}}(T_a, T_b) .$$

$\square$

6.5 Multiple Alignment

6.5.1 Feasible Lower Bound

Consider strings $S_1, \ldots, S_k$. As we have seen in Chap. 2, multiple alignment becomes easy whenever we have a tree structure on string set $S_1, \ldots, S_k$ at hand such that only pairs of string connected in the tree count for overall score (instead of all pairs as in sum-of-pairs scoring).

For this section, let us switch from score maximization to distance minimization. This is only a minor modification, but has the advantage that in terms of distance functions notions such as the triangle inequality make sense, which would be rather unusual for scoring functions. Thus we assume that we have a symmetric distance measure $d(x, y) = d(y, x)$ on pairs of characters, and $d(x, -) = d(-, x)$ on pairs of characters and spacing symbol. Further assume that $d(x, x) = 0$. Usually we also assume triangle inequality, that is $d(u, v) \leq d(u, w) + d(w, v)$. If in addition $d(u, v) > 0$ for any two different objects u and v holds, we call d a metric.

Given a distance measure d on pairs of characters and spacing symbol, we define the distance for an alignment T_1, T_2 of common length n as follows.

$$d^*(T_1, T_2) = \sum_{p=1}^{n} d(T_1(p), T_2(p)) \tag{6.13}$$

Given strings S_1 and S_2 we define their minimum distance value taken over all alignments T_1, T_2 of S_1 with S_2.

$$d_{\text{opt}}(S_1, S_2) = \min_{\text{alignments } T_1, T_2} d^*(T_1, T_2) \tag{6.14}$$

As in Sect. 6.3.3 we make use of center string S_{center} for string set $S_1, \ldots, S_k$ defined by the following equation:

$$\sum_{i=1}^{k} d_{\text{opt}}(S_{\text{center}}, S_i) = \min_{a=1,\ldots,n} \sum_{i=1}^{k} d_{\text{opt}}(S_a, S_i) \tag{6.15}$$

Lemma 6.12. *Feasible Lower Bound*
Let $T_1, \ldots, T_k$ be an optimal alignment of $S_1, \ldots, S_k$, and S_{center} be a center string for $S_1, \ldots, S_k$. Then the following holds.

$$\frac{k}{2} \sum_{i=1}^{k} d_{\text{opt}}(S_i, S_{\text{center}}) \leq d^*(T_1, \ldots, T_k)$$

Proof.

$$\frac{k}{2} \sum_{i=1}^{k} d_{\text{opt}}(S_i, S_{\text{center}})$$

$$\leq \frac{1}{2} \left(\sum_{i=1}^{k} d_{\text{opt}}(S_i, S_1) + \sum_{i=1}^{k} d_{\text{opt}}(S_i, S_2) + \ldots + \sum_{i=1}^{k} d_{\text{opt}}(S_i, S_k) \right)$$

$$= \frac{1}{2} \sum_{i=1}^{k} \sum_{j=1}^{k} d_{\text{opt}}(S_i, S_j) \leq \frac{1}{2} \sum_{i=1}^{k} \sum_{j=1}^{k} d^*(T_i, T_j) = d^*(T_1, \ldots, T_k)$$

$$\square$$

6.5.2 2-Approximation Algorithm

Lemma 6.13.
Let $T_1, \ldots, T_k$ be an optimal alignment of $S_1, \ldots, S_k$, and $C_1, \ldots, C_k$ be the alignment of $S_1, \ldots, S_k$ constructed with the tree with center string S_{center} at its root and the other strings distributed to $k - 1$ leaves as guide tree, thus for all i:

$$d^*(C_i, C_{\text{center}}) = d_{\text{opt}}(S_i, S_{\text{center}}) .$$

Then the following holds:

$$d^*(C_i, \ldots, C_k) \leq \frac{2(k-1)}{k} d^*(T_1, \ldots, T_k) .$$

Proof.

$$d^*(C_i, \ldots, C_k) = \frac{1}{2} \sum_{i=1}^{k} \sum_{j \neq i} d^*(C_i, C_j)$$

$$\leq \frac{1}{2} \sum_{i=1}^{k} \sum_{j \neq i} \left(d^*(C_i, C_{\text{center}}) + d^*(C_{\text{center}}, C_j) \right)$$

$$= \frac{1}{2}(k-1) \sum_{i=1}^{k} d^*(C_i, C_{\text{center}}) + \frac{1}{2}(k-1) \sum_{j=1}^{k} d^*(C_{\text{center}}, C_j)$$

$$= (k-1) \sum_{i=1}^{k} d^*(C_i, C_{\text{center}})$$

$$= (k-1) \sum_{i=1}^{k} d_{\text{opt}}(S_i, S_{\text{center}})$$

$$\leq \frac{2(k-1)}{k} d^*(T_1, \ldots, T_k)$$

$$\square$$

6.6 Sorting Unsigned Genomes

Given a permutation $\pi(1), \ldots, \pi(n)$ of numbers $1, \ldots, n$, *undirected reversal* between indices i, j with $i \leq j$ inverts order of numbers $\pi(i), \ldots, \pi(j)$. The least number $d(\pi)$ required to lead to the sorted permutation $1, \ldots, n$ is called the undirected reversal distance of permutation π. As was mentioned in Chap. 5, computation of the undirected reversal distance for arbitrary permutations is NP-hard.

The introduction of boundary numbers 0 and $n + 1$ (comparable to node L and R in an RD-diagram) simplifies some definitions (which without these numbers would lead to a steadily distinction of whether we are working inside a permutation, or at the left or right border). Thus we are from now on working with $0, \pi(1), \ldots, \pi(n), n + 1$ as initial and $0, 1, \ldots, n, n + 1$ as final permutation. However, note that reversals always take place between some left position > 0 and right position $< n + 1$ (thus 0 and $n + 1$ are never moved).

6.6.1 Feasible Lower Bound

If in a permutation $0, \pi(1), \ldots, \pi(n), n+1$ two numbers a and b with $|a - b| = 1$ occur as neighbours, there is no urgent necessity to execute a reversal separating them. The reason is that "locally" at that position the permutation is already sorted. Of course, it might nevertheless be advantageous to let a reversal temporarily separate both numbers. Situation is different at all positions between two neighbouring numbers a and b for which $|a - b| > 1$ holds. Any such position is called a *breakpoint* of permutation $0, \pi(1), \ldots, \pi(n), n+1$. At any breakpoint we are forced to apply a reversal that separates the number immediately left of the breakpoint from the number immediately right of the breakpoint. We denote by $b(\pi)$ the number of breakpoints of permutation $0, \pi(1), \ldots, \pi(n), n + 1$. As an example, permutation $0, 7, 2, 3, 9, 6, 5, 4, 1, 8, 10$ has 10 positions, 7 of them being breakpoints.

We require some more notions. Given a permutation $0, \pi(1), \ldots, \pi(n), n+1$, we call a contiguous segment of maximum length that does not contain a breakpoint a *block*. Blocks that do not contain 0 or $n + 1$ are called *inner blocks*. Every block either consists of a monotonically increasing sequence of numbers, called an *increasing block*, or of a monotonically decreasing sequence of numbers, called a *decreasing block*. As an example, the permutation in Fig. 6.7 is segmented into increasing (arrow upwards) and decreasing (arrow downwards) blocks as follows (blocks of length 1 are at the same time increasing and decreasing).

Lemma 6.14. *Feasible Lower Bound*

$$\frac{1}{2}b(\pi) \leq d(\pi)$$

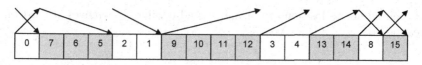

Fig. 6.7. Increasing and decreasing blocks

Proof. Any reversal may reduce at most two breakpoints (at its borders). Thus, at least $1/2\,b(\pi)$ reversals are required to sort the permutation. □

6.6.2 4-Approximation Algorithm

Lemma 6.15.
Whenever there is an inner decreasing block, there exists a reversal that decreases $b(\pi)$ by 1.

Proof. Assume that there is at least one inner decreasing block. Let m be the minimum of all numbers within inner decreasing blocks, and i be its position within the permutation. Thus we know that $1 \le m \le n$ and $1 \le i \le n$, as only inner blocks are considered. Now we also look at number $m-1$ and its position j within the permutation. Note that $m-1=0$ is possible. We collect some information about numbers m and $m-1$.

- Number m is the rightmost number of its inner decreasing blocks since otherwise $m-1$ would be a smaller number within an inner decreasing block.
- As number $m-1$ is not located within an inner decreasing block, $m-1$ must be located either within an inner increasing block, or in the (always increasing) block of number 0, or in the (always increasing) block of number $n+1$. Either way, $m-1$ is located within an increasing block.
- Numbers m and $m-1$ cannot occur immediately consecutive in that order since otherwise also $m-1$ would be part of the inner decreasing block of m.
- Numbers $m-1$ and m cannot occur immediately consecutive in that order since otherwise m would not be part of an inner increasing block.
- Numbers $m-1$ and m are separated by at least one number.
- Number $m-1$ is the rightmost number within its increasing block since otherwise next number m would follow immediately right of $m-1$.

Now we consider the positions i and j of number m and $m-1$. First look at the case $i < j$. The situation is as shown in Fig. 6.8 with breakpoints indicated by arrows. A broken arrow indicates a position that might still be a breakpoint after the reversal, but also might have been removed as a breakpoint. Number $m-1$ cannot be $n+1$, thus the reversal between positions $i+1$ and j is admissible. It removes at least one breakpoint. Second consider the case $j < i$. Situation is as shown in Fig. 6.9. Number m cannot be $n+1$, thus the reversal between numbers positions $j + 1$ and i is admissible. It removes at least one breakpoint.

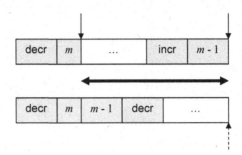

Fig. 6.8. Removal of a breakpoint

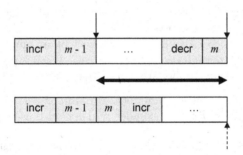

Fig. 6.9. Removal of a breakpoint

□

Lemma 6.16.
If $0, \pi(1), \ldots, \pi(n), n + 1$ is a permutation different from the identity permutation and without an inner decreasing block, then there exists at least one inner increasing block. Application of a reversal that inverts such an inner increasing block generates an inner decreasing block and does not increase the number of breakpoints.

Proof. Consider the block B that contains number 0. This cannot be all of $0, \pi(1), \ldots, \pi(n), n + 1$ since otherwise we had the identity permutation. Also consider block B' that contains number $n + 1$. B and B' are different blocks, and both are increasing blocks (eventually at the same time also decreasing blocks). Between B and B' there must exist at least one further block B'' since otherwise the permutation would be the identity permutation. B'' is an inner block, thus by assumption increasing. The rest of the lemma is clear.   □

Theorem 6.17.

Below we present an efficient algorithm which reverses an arbitrary inner decreasing block if present, and thus reduces $b(\pi)$ by 1 (Lemma 6.15), or inverts an arbitrary inner increasing block otherwise (Lemma 6.16). As in every second step or earlier number $b(\pi)$ is decreased by 1, the algorithm requires at most $2b(\pi)$ reversals to sort a given permutation with $b(\pi)$ breakpoints. Using the already shown estimation $1/2\,b(\pi) \leq d(\pi)$ we conclude that the algorithm uses at most 4 times as many reversals as possible in the optimum case. Thus we have a 4-approximation algorithm.

```
while π ≠ identity permutation do
    if there is no inner decreasing block then
        invert an arbitrary inner increasing block
    else
        m = minimum number in all inner decreasing blocks
        let i be the position of m
        let j be the position of m − 1
        if i < j then
            reverse between i + 1 and j
        else
            reverse between j + 1 and i
        end if
    end if
end while
```

6.6.3 2-Approximation Algorithm

A slightly more careful analysis improves the approximation factor from 4 to 2.

Lemma 6.18.

Assume that $0, \pi(1), \ldots, \pi(n), n + 1$ contains an inner decreasing block. Then a reversal can be chosen that either decreases $b(\pi)$ by 1 and leads to a permutation which again contains an inner decreasing block, or decreases $b(\pi)$ by 2.

In the former case, we may proceed to decrease $b(\pi)$ without need to provide for an inner decreasing block by an extra reversal. In the latter case, reduction of $b(\pi)$ by 2 allows to execute a required extra reversal that creates an inner decreasing block. Thus we obtain the following conclusion.

Corollary 6.19.

There is a 2-approximation algorithm for the unsigned genome rearrangement problem.

Proof. Assume that $0, \pi(1), \ldots, \pi(n), n+1$ contains an inner decreasing block. As before, we consider the minimum number m occurring in inner decreasing blocks, as well as number $m - 1$; let i be the position of m and j be the position of $m - 1$. In case $i < j$ (see Fig. 6.8) the considered reversal between $i + 1$ and j again produced an inner decreasing block, namely the block that contains m (observe that this is indeed an inner block). In case that $j < i$ holds, things get more difficult. In particular, Fig. 6.9 does not immediately show, after the reversal between $j + 1$ and i, an inner decreasing block. Assume that there is no reversal that decreases $b(\pi)$ by 1 and leads to a permutation which again contains an inner decreasing block. Now we also consider the maximum number M occurring in inner decreasing blocks, and let r be its position. Finally consider number $M + 1$, and let s be its position. We collect what we know about numbers m, $m - 1$, M, $M + 1$, and their positions:

- Number m is the rightmost element within its inner decreasing block.
- Number $m - 1$ is the rightmost element within its increasing block.
- Number M is the leftmost element within its inner decreasing block.
- $M < n + 1$ as $n + 1$ does not belong to an inner block.
- Number $M + 1$ is the leftmost element within its increasing block.
- Numbers m and $m - 1$ are not at neighbouring positions in the permutation.
- Number M cannot be immediately left of $M + 1$ since then M would be part of an increasing block.
- Number $M + 1$ cannot be immediately left of M since then M would not be the maximum number within inner decreasing blocks.
- Positions i and r are between 1 and n, position j may be 0, position s may be $n + 1$.

In the following, positions i, j, r, s are more precisely determined. In the figures shown below, breakpoints are always indicated by arrows. A broken arrow always indicates a position where a breakpoint may still be present, but also may have disappeared after the reversal.

- $r < s$

In case that $s < r$, the reversal between s and $r - 1$ would decrease $b(\pi)$ by 1 and lead again to an inner decreasing block (the block containing number M; see Fig. 6.10), contrary to our assumption.

- $r \leq i$

In case that $i < r$, the reversal between $j + 1$ and i would decrease $b(\pi)$ by 1 and lead again to an inner decreasing block (the inner block containing number M; see Fig. 6.11), contrary to our assumption.

- $j < r$

In case that $r \leq j$, the reversal between $j + 1$ and i would decrease $b(\pi)$ by 1 and lead again to an inner decreasing block (the inner block containing number M; see Fig. 6.12), contrary to our assumption.

Thus, at the moment we know that $r < s$ und $j < r \leq i$ (Fig. 6.13).

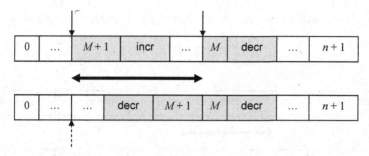

Fig. 6.10. Breakpoint removal

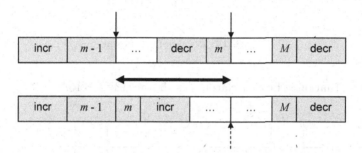

Fig. 6.11. Breakpoint removal

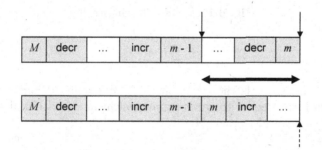

Fig. 6.12. Breakpoint removal

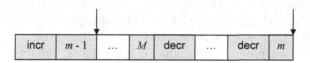

Fig. 6.13. Interim result

- $i \leq s$

In case that $s < i$ we had $r < s < i$, thus the reversal between r and $s - 1$ would decrease $b(\pi)$ by 1 and lead again to an inner decreasing block (the inner block containing number m; see Fig. 6.14), contrary to our assumption.

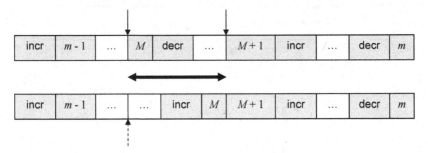

Fig. 6.14. Breakpoint removal

Thus the situation is further clarified as shown in Fig. 6.15.

Fig. 6.15. Refined interim result

- $i < s$

This is obvious, as m is the minimum element within all inner decreasing blocks, thus less or equal to the maximum element M within all inner decreasing blocks, thus less than $M + 1$.

- $i + 1 = s$

Assume that $i + 1 < s$ holds. Then there would be a further inner block B between the block of m and the block of $M + 1$. In case that B is decreasing, the reversal between $j + 1$ (right of $m - 1$) and i (position of m) would remove one breakpoint and let B invariant as inner decreasing block, contrary to our general assumption. In case that B is increasing, the reversal between r (position of M) and $s - 1$ (left of $M + 1$) would remove one breakpoint and create an inner decreasing block from B, contrary to our general assumption.

Thus situation is further clarified as shown in Fig. 6.16.

- $j + 1 = r$

Assume that $j + 1 < r$ holds. Then there would be a further inner block B between the block of $m - 1$ and the block of M. In case that B is decreasing,

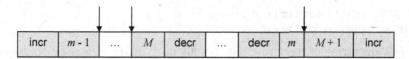

Fig. 6.16. Refined interim result

the reversal between r (position of M) and $s-1$ (left of $M-1$) would remove one breakpoint and let B invariant as inner decreasing block, contrary to our general assumption. In case that B is increasing, the reversal between $j+1$ (right of $m-1$) and i (position of m) would remove one breakpoint and create an inner decreasing block from B, contrary to our general assumption.

Thus situation is further clarified as shown in Fig. 6.17. Application of the reversal between r (position of M) and i (position of m) at once removes two breakpoints.   □

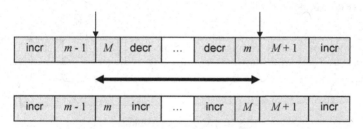

Fig. 6.17. Removal of two breakpoints

6.7 HP-Model Protein Structure Prediction

6.7.1 Feasible Upper Bound

Lemma 6.20.
Whenever in an HP-fold of string S two bits are in contact then there must be an even number of bits between them.

Proof. Obvious? If not, take a look at the fold in Fig. 6.19.   □

As we are interested in contacts between bits 1, it plays a role whether two occurrences of bit 1 are separated by an even or odd number of zeroes. This leads to a certain decomposition of any bit string that plays a crucial role in the following. Following the conventions with regular expressions we denote by:

- $(00)^*$ an even number of zeroes

- (00)*0 an odd number of at least one zero
- 0* an arbitrary number of zeroes

We call any string of the form $1(00)^*01(00)^*01\ldots1(00)^*01$ a *block*. Note that a block starts and ends with bit 1, each of its bits 1 is separated from the next bit 1 through an odd number of bits 0. A block may consist of bit 1 alone. Now any bit string S can be uniquely decomposed into:

- a prefix consisting of an arbitrary number of bits 0
- blocks that are separated from the next block by an even number (can be zero) of bits 0
- a suffix consisting of an arbitrary number of bits 0

We denote in an alternating manner blocks as X-blocks X_i and Y-blocks Y_i. The sense of this alternation will soon become clear.

$$S = 0^*X_1(00)^*Y_1(00)^*X_2(00)^*Y_2(00)^*X_3(00)^*Y_3(00)^*\ldots(00)^*X_k/Y_k0^*$$

As an example consider the string in Fig. 6.18. Block structure is indicated by shadowed areas.

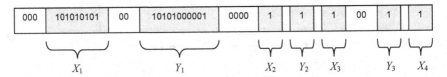

Fig. 6.18. X- and Y-blocks

Lemma 6.21.
Given two bits 1 in the same or in different X-blocks, there is an odd number of bits between them. The same holds for bits 1 within the same or different Y-blocks. Thus, in an HP-fold, there can be contacts between bits 1 only in case that one is located in an X-block, and the other one is located in a Y-block.

Proof. Convince yourself from the correctness of the lemma by inspecting various combinations of two bits 1 in the example of Fig. 6.18. Alternatively, give a formal proof. □

For an arbitrary string S consider its unique block structure as defined above and count the number $x(S)$ of bits 1 in its X-blocks, and the number $y(S)$ of bits 1 in its Y-blocks. For simplicity, assume from now on that $x(S)$ and $y(S)$ are even numbers (and think about necessary modifications in the following for the case that one or two of these numbers are odd).

Lemma 6.22.
For a string of length n with numbers $x(S)$ and $y(S)$ as defined above, any HP-fold of S can contain at most $2\min\{x(S), y(S)\} + 2$ contacts between two bits 1.

Proof. Assume that $x(S) \leq y(S)$. In every contact between two bits 1 there must be exactly one bit used from an X-block. Every node occupied by bit S_i in the HP-model can make at most two contacts with other bits as one of its four neighbour nodes is occupied by the immediate predecessor bit S_{i-1}, and another one of its four neighbour nodes is occupied by the immediate successor bit S_{i+1} (these both do not count as HP-contacts). This rule has two exceptions: nodes that are occupied by S_1 and S_n may have at most three contacts. In summary, there are at most $2x(S) + 2$ many contacts in any HP-fold of S.   □

6.7.2 Asymptotic 4-Approximation Algorithm

Lemma 6.23. *Split Lemma*

Every string S can be decomposed into $S = S'S''$ such that either S' contains at least 50% of all bits 1 occurring in X-blocks of S and S'' contains at least 50% of all bits 1 in Y-blocks of S, or conversely S'' contains at least 50% of all bits 1 occurring in X-blocks of S and S' contains at least 50% of all bits 1 occurring in Y-blocks of S (it does not matter where string S is separated; split position might as well be within an X- or Y-block without causing problems in the construction of a fold below).

Proof. Traverse S from left to right and determine the leftmost position p in S such that left of position p there are exactly 50% of the bits 1 occurring in X-blocks of S. By the same way, determine the rightmost position q in S such that right of q there are exactly 50% of the bits 1 occurring in Y-blocks of S (think about what has to be modified in case that $x(S)$ and/or $y(S)$ are not even). In case that $p \leq q$ we may split S anywhere between p and q and obtain the first variant of the conclusion of the lemma, in case that $q < p$ we split S anywhere between q and p and obtain the second variant of the conclusion of the lemma.   □

From now on let us treat the case $S = S'S''$ with S' containing at least 50% of the bits 1 occurring in X-blocks of S, and S'' containing at least 50% of bits 1 occurring in X-blocks of S. We now fold S' and S'' in a very special manner, placing all of the bits 1 in S' that belong to an X-block of S above a "borderline", and all bits 1 in S'' that belong to a Y-block of S below this borderline such that as many as possible of them are in contact. Details are first made clear with an chosen example, and then the construction is described more generally.

As an extended example, we treat $S = S'S''$ decomposed as shown below. Note that $x(S) = 12$ and $y(S) = 14$. S' contains 6 of the 12 bits 1 of X-blocks in S, and S'' contains 10 of the 14 bits 1 in Y-blocks of S.

$$S' = 000\ \underbrace{1010000010001}_{X_1}\ 00\ \underbrace{1010001}_{Y_1}\ \underbrace{1}_{X_2}\ \underbrace{1}_{Y_2}\ \underbrace{1}_{X_3}\ 0000$$

$$S'' = \underbrace{10000010100010001}_{Y_3}\ 0000\ \underbrace{10001010100000010001}_{X_4}\ 00\ \underbrace{10100000101000001}_{Y_4}$$

The constructed fold is shown in Fig. 6.19. Bits 1 in X-blocks are drawn as black dots, bits 1 in Y-blocks as grey dots, and bits 0 as white dots. Note that black dots from S' occur above the borderline at every second position, with a white or grey dot separating successive black dots. At the right end of the borderline we bend the fold in such a way that the leftmost 1 from S'' is below the rightmost 1 from S'. Now the same is done for S'' below the borderline that was done for S' above the borderline: grey dots from S'' occur below the borderline at every second position, with a white or black dot separating successive grey dots. By this way, at least $1/2 \min\{x(S), y(S)\}$ pairings between black and grey dots are realized.

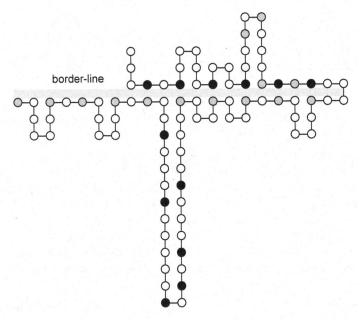

Fig. 6.19. Special fold along borderline used by the 4-approximation algorithm

Figure 6.20 again explains the construction for the general case. The number of nodes between any two consecutive bits 1 in an X-block is odd. Also, the number of nodes between last bit 1 of an X-block and first bit 1 of the next X-block is odd as the Y-block between has an odd number of nodes. This makes it possible to place all bits 1 of X-blocks of S' at every second position of the indicated borderline. As the number of nodes between last 1 in S' belonging to an X-block and first 1 in S'' belonging to an Y-block is even, it is possible to place this first 1 of S'' on the borderline exactly below this last 1 of S'. Now the game proceeds with Y-blocks of S'' below the borderline exactly the same way as with X-blocks of S' above the borderline. As a result, a number of $1/2 \min\{x(S), y(S)\}$ many pairs of black and grey nodes can be placed on the borderline side-by-side (though eventually one of these pairs of black and grey nodes does not count as a contact, as is made clear below).

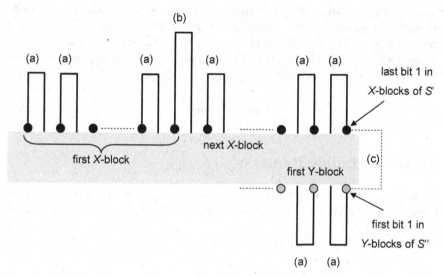

Fig. 6.20. Black nodes represent all bits 1 of X-blocks of string S'. Each of the paths starting above a black node on the borderline consists of an odd number of nodes and ends with a node on the borderline. Bold lines **(a)** indicate blocks of all zeroes that separate consecutive bits 1 within X-blocks; broken lines **(b)** indicate blocks consisting of all zeroes between an X-block and next Y-block together with the complete next Y-block; "turn around" **(c)** consists of an even number of nodes between black and grey dot. Below borderline the same picture results with Y-blocks of string S'' (bits 1 of Y-blocks within S'' are represented by grey dots).

As Fig. 6.19 for the particular example shows, and Fig. 6.20 makes clear for the general case, we manage to create a fold with a number of contacts of bits 1 that is at least $\min\{1/2\,x(S), 1/2\,x(S)\} - 1$. Note where the '-1' comes from. It might be the case that the last 1 in the last X-block of S' is immediately followed by the first 1 in the first Y-block of S'' (for example, think of strings consisting only of bits 1). In that case, this single pair of immediately consecutive bits 1 does not count as a contact. Now we put this number in relation to the maximum possible number max of folds as follows:

$$\text{max} \leq 2\min\{x(S), y(S)\} + 2$$

$$\min\{x(S), y(S)\} \geq \frac{1}{2}(\text{max} - 2) \tag{6.16}$$

$$\text{folds} \geq \frac{1}{2}\min\{x(S), y(S)\} - 1 \geq \frac{1}{2}\frac{1}{2}(\text{max} - 2) - 1 = \frac{1}{4}\text{max} - \frac{3}{2}$$

This looks almost as expected for a 4-approximation algorithm, if we ignore the constant additive term $-1\frac{1}{2}$. Indeed, it is an example of a 4-approximation algorithm *in the asymptotic sense*. We omit a formal definition of what "asymptotic sense" precisely means (it is easy and can be found in every textbook on complexity theory), but explain it informally: as string S gets larger and larger and, related to this, also the maximum number of contacts in any HP-fold of S grows, constant additive term $-1\frac{1}{2}$ becomes more and more negligible. Suppressing it at all gives the estimation for the case of a-approximation (note that factor $1/4$ for maximization corresponds to factor 4 for minimization problems).

$$\text{folds} \geq \frac{1}{4}\max \qquad (6.17)$$

6.8 Bibliographic Remarks

Good introductions to the theory of approximation algorithms can be found in Papadimitriou [61], a more extended treatment can be found in the textbook of Vazirani [74].

A Selection of Metaheuristics and Various Projects

Besides the algorithm design paradigms presented in the preceding chapters that have a clear theoretical foundation, various more or less well understood heuristics are applied to solve bioinformatics problems, often with astonishing good results. Though emphasis of this textbook is clearly on strict formal methods, these heuristics should not be completely omitted. Instead of talking of "heuristics" we use the term "metaheuristics" here, as the presented methods are more universal frameworks for the solution of various problems, and not so much single heuristics suited for the solution of one, and only one problem. Without any attempts to be complete, the following metaheuristics are presented and illustrated with the aid of some projects. The reader interested in getting a deeper understanding of one of these methods may then consult one of the recommended textbooks. Each of these attempts is briefly presented only as far as it is required for the understanding of the treated applications.

- Multi-layer perceptrons [4, 12]
- Support vector machines [24]
- Hidden Markov models [4, 46]
- Ant colony optimization [25]
- Genetic algorithms [54, 72]

7.1 Multi-Layer Perceptron

We might take the standpoint that we use neural networks as black boxes without asking what happens inside the box. Up to some extend, one may indeed hope that an efficient neural network simulator does a good job and we only have to take care of the proper design of the interface between the application problem and the black box. As it is well known from many projects, proper design of interfaces is at the heart of a successful project implementation on basis of neural networks. This is often stated as the problem of selecting features that appropriately encode problem input. To get some feeling why this is so, and to also get some indication of what proper feature selection might mean, it is helpful to have some (fortunately limited) insights into the inner machinery of neural networks. The following sections try to present such a very comprehensive crash course in neural network machinery.

7.1.1 Architecture of Multi-Layer Perceptrons

A single neuron is a device that receives real valued *signals* $x_1, \ldots, x_n$, either from predecessor neurons or from input interface, over wires that have attached *weights* $w_1, \ldots, w_n$. It integrates weighted input signals together with its *bias b* into a sum called *net input*. An *activation function f* then determines what value $y = f(\text{net})$ is returned as output signal (either as input signal for further neurons or returned as output signal).

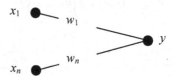

Fig. 7.1. Signal neuron receiving weighted inputs

Various activation functions are common as, for example, *sigmoid function* or *thresholds function*. Formal details are as follows.

$$\text{net} = \text{net}(w_1, \ldots, w_n, x_1, \ldots, x_n, b) = \sum_{i=1}^{n} w_i x_i + b$$

$$y = f(\text{net})$$

$$\text{sigmoid funtion } f(\text{net}) = \frac{1}{1 + e^{-\text{net}}}$$

$$\text{thresholds function } f(\text{net}) = \begin{cases} 1 & \text{net} > 0 \\ 0 & \text{net} \leq 0 \end{cases}$$

Now a *multi-layer perceptron* consists of an ensemble of neurons that are arranged into several layers such that there are only connections from each layer to the next one right of it. Leftmost layer is called the *input layer*, rightmost layer is called the *output layer*, and any further layer between these two layers is called a *hidden layer*. Figure 7.2 shows an example of a multi-layer perceptron with 6 neurons in its input layer, 4 neurons in its first and 2 neurons in its second hidden layer, and a single output neuron.

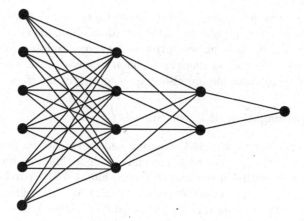

Fig. 7.2. Example multi-layer perceptron

7.1.2 Forward Propagation of Input Signals

Given input signals $x_1, \ldots, x_n$, these are propagated through the layers of a network from left to right according to the following formula which defines for each neuron i its activation value y_i assuming that y_j is already computed for all predecessor neurons j of neuron i (in the formula, w_{ji} denotes the weight of the link leading from predecessor neuron j to neuron i):

$$y_i = \begin{cases} x_i & \text{for input neuron } i, \\ f\left(\displaystyle\sum_{\substack{\forall j \text{ with} \\ j \to i}} w_{ji} y_j + b_i \right) & \text{otherwise.} \end{cases}$$

Thus, any multi-layer perceptron with n input neurons and m output neurons defines a real-valued vector-function $f : \mathbb{R}^n \to [0,1]^m$ such that $f(x_1, \ldots, x_n)$ are returned as activation values $y_1, \ldots, y_m$ which occur at the output neurons after propagation of input through the net. For the case of thresholds network we obtain a bit-valued vector-function $f : \mathbb{R}^n \to \{0,1\}^m$.

7.1.3 Input and Output Encoding

In case of binary classification problems the design of input and output interfaces of a multi-layer perceptron is obvious. In case that we have to deal with strings over more complex alphabets, for example the 20 amino acid alphabet of proteins, there are various ways of encoding characters as signals for a multi-layer perceptron. For the example of 20 characters, one way is to use a single input neuron for each character that gets a real valued signal from interval $[0,1]$ partitioned into 20 segments of equal size, or to use 5 neurons for each character with a binary encoding expressing which one of the 20 characters is presented as input, or to use 20 neurons with an unary encoding (exactly one neuron gets value 1, the other ones get value 0) expressing which one of the 20 characters is presented as input. Most often, unary encoding give best results, as the multi-layer perceptron is not forced to learn, in addition to the proper task, also the peculiarities of binary numbers. All what has been said about input interface design also applies to output interface design.

7.1.4 Backpropagation of Error Signals

Multi-layer perceptrons attracted attention and experienced a renaissance in the 1980s, when the famous backpropagation algorithm was invented that provided the field with a mathematically well-founded and particularly simple to implement learning rule. We consider multi-layer perceptrons with the sigmoid function as activation function (as a differentiable activation function is required for the gradient descent based backpropagation procedure). Being a gradient descent procedure that attempts to minimize quadratic error observed at the output layer, its formula structure is particularly simple (and particularly easily derived using the chain rule for partial derivatives of differentiable functions of several variables) and is almost perfectly similar to the forward propagation of activation values. After presenting an input vector x to the network, propagating it through the network with a computation of net inputs net_i and output values y_i for every neuron i, and expecting output vector d at the output layer, weights w_{ij} are updated on basis of a quadratic error function gradient descent procedure by values Δw_{ij} that are computed in a backward propagation from the output layer back to the input layer (with a learning rate $\eta > 0$). Though defined as partial derivatives, by the use of the chain rule for partial derivatives formulas reduce to simple sums. This simple structure of backpropagation makes it particularly attractive. Formal details are as follows, with w being the vector of all weights (to make the formulas easy to read, dependence on w, x, and d are not explicitly indicated as variables in the functions occurring below).

$$\text{error} = \frac{1}{2} \sum_{\substack{\text{output} \\ \text{neuron } i}} (y_i - d_i)^2$$

$$\frac{\partial \text{error}}{\partial w_{ij}} = \frac{\partial \text{error}}{\partial \text{net}_j} \frac{\partial \text{net}_j}{\partial w_{ij}} = \delta_j y_i$$

Here, we used the abbreviation $\delta_j = \frac{\partial \text{error}}{\partial \text{net}_j}$. Terms δ_j are computed from output layer back to input layer using a second time chain rule. For output neuron j we obtain[1]:

$$\delta_j = \frac{\partial \text{error}}{\partial y_j} \frac{\partial y_j}{\partial \text{net}_j} = (y_j - d_j)f'(\text{net}_j) = (y_j - d_j)y_j(1 - y_j).$$

---

[1] Derivative of sigmoid function f is computed as $f'(\text{net}) = f(\text{net})(1 - f(\text{net}))$ (easy exercise).

For neuron j that is not an output neuron we assume that δ_k is already computed for all successor neurons k of neuron j. Then we obtain:

$$\delta_j = \sum_{\substack{\text{neurons } k \\ \text{with } j \to k}} \frac{\partial \text{error}}{\partial \text{net}_k} \frac{\partial \text{net}_k}{\partial y_j} \frac{\partial y_j}{\partial \text{net}_j}$$

$$= \sum_{\substack{\text{neurons } k \\ \text{with } j \to k}} \delta_k w_{jk} f'(\text{net}_j)$$

$$= y_j(1 - y_j) \sum_{\substack{\text{neurons } k \\ \text{with } j \to k}} w_{jk} \delta_k$$

Summarizing, backpropagation comprises the following gradient descent update rule for weights $\Delta w_{ij} = -\eta y_i \delta_j$ with:

$$\delta_j = \begin{cases} (y_j - d_j) y_j (1 - y_j) & \text{for output neuron } j, \\ y_j(1 - y_j) \sum_{j \to k} w_{jk} \delta_k & \text{otherwise.} \end{cases}$$

7.1.5 Tuning Backpropagation

To turn backpropagation into a practically working learning rule requires some more effort; we mention a few key words: avoiding stagnation of error minimization within flat plateaus of error landscape; avoiding oscillation in narrow valleys; momentum term; dynamic learning rate adaptation; flat spot elimination; early stopping; weight decay; optimal brain damage; architecture optimization by genetic algorithms; presentation schedule for training examples. The reader interested in details may consult any standard textbook on neural networks.

7.1.6 Generalization and Overfitting

Looking at multi-layer perceptrons as black boxes, we must not so much worry about such technical details: hopefully an intelligent simulator does this job for us. Much more important is to be aware of the steady danger of overfitting due to too many degrees of freedom within a network, and the importance for a very clever selection of features that are presented as input to a network. As an example, learning to classify photographic images of human faces into "male" or "female" does not make sense with a pixel representation of faces. Instead one should think of features as hair style, lip colouring, and the like, as input signals for the network. In statistical learning theory, we attempt to achieve the following situation. With 99% probability (as an example value) over a randomly drawn (with respect to an arbitrary, unknown distribution of training data) training set T of size m, we want to have an upper bound for the generalization error that may occur in the worst case after having chosen a classification function h from some function (hypothesis) space H that classifies training data correctly. Only for restricted such function spaces one may hope to get such an estimation for the generalization error. Using as function space the set of all functions realized by a multi-layer perceptron with a number W of adjustable parameters (weights and biases), and a number N of neurons, the following rough estimations (omitting any constants, as well as some minor factors) for generalization error are known in statistical learning theory:

1. In case of threshold neurons, generalization error roughly scales with $\frac{1}{m}W \log(W)$.
2. In case of sigmoid neurons, generalization error roughly scales with $\frac{1}{m}W^2 N^2$.

Though being rather imprecisely stated, these estimations nevertheless shed light on generalization ability of neural networks: on the one side, the more training data is available the higher is the chance of good generalization. On the other side, with increasing size of the network, expected generalization error increases too. In particular, input dimension influences generalization error for a multi-layer perceptron. This explains why it is so essential to properly design input interface to a neural network: using appropriate features is at the heart for obtaining a well-working network. Besides careful (and sparse) usage of input features, pruning techniques that reduce the number of weights and neurons as much as possible (without compromising training success) support generalization ability.

7.1.7 Application to Secondary Structure Prediction

In Sect. 2.10.1 we discussed the problem of predicting the secondary structure class of an amino acid in the middle of a window of width 13 amino acids. Having available a large number of classified such windows, the problem is ideally suited for a multi-layer perceptron approach. Use 13×20 input neurons to represent a window of 13 amino acids in unary manner, and 3 output neurons to represent the class 'alpha helix', 'beta sheet', or 'loop' of the centre amino acid in a unary manner. With a limited number of hidden neurons within two hidden layers it is possible to obtain good prediction results (see [65]). Using, for example, 20 neurons in the first hidden layer, and 5 neurons in the second hidden layer, we obtain a neural network with an overall number of $260 \times 20 + 20 \times 5 + 5 \times 3 = 5215$ weights. Thus architecture of a multi-layer perceptron looks as shown in Fig. 7.3.

7.1.8 Further Applications

The most comprehensive overview of neural network applications to bioinformatics is surely Baldi & Brunak [4]. We mention a few applications that are treated in this book, but do not go into more details of the problems as the

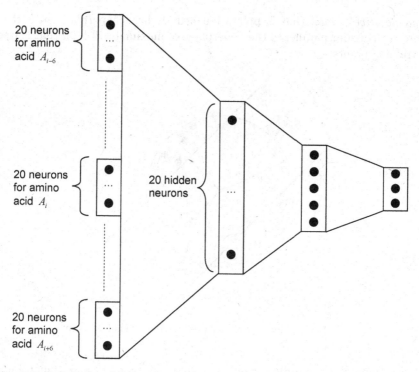

Fig. 7.3. Standard architecture of a multi-layer perceptron for secondary structure prediction

principled way to apply multi-layer perceptrons is always the same along the lines presented with the example above. What differs and is problem-specific is the proper design of interfaces to some neural network architecture and, of course, the selection of suitable network architecture. Applications deal with the identification and prediction of, for example:

- Intron splice sites [14]
- Signal peptides [57]
- Cleavage sites [57]
- Beta-sheet prediction [49]

7.2 Support Vector Machines

7.2.1 Architecture of Support Vector Machines

A *training set* T consists of m labelled points p located in a ball of radius R (radius R is required later in an estimation of generalization error) in n-dimensional real space, and associated labels $d \in \{+1, -1\}$. Support vector machines aim to linearly separate *positive points*, that is points labelled $+1$, from *negative points*, that is points labelled -1, by a hyperplane such that there is a corridor parallel to the hyperplane of maximal width 2μ that is free of training points.

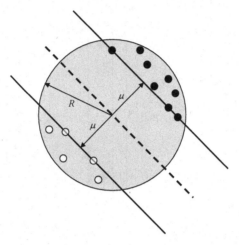

Fig. 7.4. Corridor separating positive from negative points

Any hyperplane in n-dimensional real space is defined by an n-dimensional direction vector w (perpendicular to hyperplane) and a bias (offset) b as follows:

$$H(w, b) = \{x \in \mathbb{R}^n \mid \langle w, x \rangle + b = 0\}$$

Here $\langle w, x \rangle$ denotes inner product of vector w with vector x. Decomposing any vector p into its component $p_{=}$ parallel to vector w, and $p_{\perp}$ orthogonal to w, we easily compute that $H(w, b)$ has signed distance (positive, zero, or negative) $b\|w\|^{-1}$ to point $(0,0)$, and signed distance (positive, zero, or negative) $(\langle w, p \rangle + b)\|w\|^{-1}$ to an arbitrary point p.

Exercise

7.1. Verify the formulas above.

7.2.2 Margin Maximization

Consider training set $T = (x^1, d^1), \ldots, (x^m, d^m)$. Separation of positive and negative points by hyperplane $H(w, b)$ means:

$$d^i = +1 \rightarrow \langle w, x^i \rangle + b > 0$$

$$d^i = -1 \rightarrow \langle w, x^i \rangle + b < 0.$$

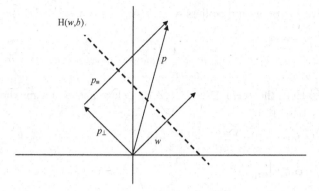

Fig. 7.5. Computation of distance of point p from hyperplane $H(w, b)$

By rescaling weight vector w and bias b (that does not alter the hyperplane considered) we may normalize the requirements above always as follows:

$$d^i = +1 \rightarrow \langle w, x^i \rangle + b \geq +1$$

$$d^i = -1 \rightarrow \langle w, x^i \rangle + b \leq -1.$$

Both requirements can be subsumed by the following single one:

$$d^i(\langle w, x^i \rangle + b) - 1 \geq 0.$$

Now one is interested in a hyperplane $H(w, b)$ that maximizes the distance of training points from the hyperplane, called *margin*:

$$\mu_T(w, b) = \min_{i=1,\ldots,m} \frac{|\langle w, x^i \rangle + b|}{\|w\|}.$$

7.2.3 Primal Problem

The discussion in Sect. 7.2.3 leads to the following *primal* margin maximization problem:

MARGIN (primal form)
Given training set $T = (x^1, d^1), \ldots, (x^m, d^m)$,
minimize $\|w\|$ subject to $d^i(\langle w, x^i \rangle + b) - 1 \geq 0$.

Note that weight vector as well as bias has to be found. The margin maximization problem is a quadratic optimization problem with linear constraints. As for every real valued optimization problem with real valued constraints, the general technique of *Lagrange multipliers* tells how to solve such a problem. We do not introduce here to the theory of Lagrange multipliers, but directly present what sort of problem finally results from the margin maximization problem. For the following function depending on $w_1, \ldots, w_n$, b, and non-negative Lagrange multipliers $\alpha_1, \ldots, \alpha_m$ a so-called *saddle-point* has to be found:

$$L(w_1, \ldots, w_n, \alpha_1, \ldots, \alpha_m, b) = \frac{1}{2}\langle w, w \rangle - \sum_{i=1}^{m} \alpha_i(d^i(\langle w, x^i \rangle + b) - 1).$$

This means that the term $Q(\alpha_1, \ldots, \alpha_m)$ below has to be maximized over $\alpha_1 \geq 0, \ldots, \alpha_m \geq 0$:

$$Q(\alpha_1, \ldots, \alpha_m) = \inf_{w_1, \ldots, w_n, b} L(w_1, \ldots, w_n, \alpha_1, \ldots, \alpha_m, b).$$

7.2.4 Dual Problem

To compute $Q(\alpha_1, \ldots, \alpha_m)$, for fixed values of $\alpha_1, \ldots, \alpha_m$, partial derivatives of function L with respect to w_k and b have to be computed and set to zero:

$$0 = \frac{\partial L}{\partial w_k} = w_k - \sum_{i=1}^{m} \alpha_i d^i x_k^i$$

$$0 = \frac{\partial L}{\partial b} = -\sum_{i=1}^{m} \alpha_i d^i.$$

Thus we get representations:

$$w = \sum_{i=1}^{m} \alpha_i d^i x^i$$

$$\sum_{i=1}^{m} \alpha_i d^i = 0.$$

Introducing the w obtained into the definition of function L yields:

$$L(w_1, \ldots, w_n, \alpha_1, \ldots, \alpha_m, b)$$
$$= \frac{1}{2} \sum_{i=1}^{m} \sum_{j=1}^{m} \alpha_i \alpha_j d^i d^j \langle x^i, x^j \rangle - \sum_{i=1}^{n} \sum_{j=1}^{n} \alpha_i d^i \alpha_j d^j \langle x^i, x^j \rangle + \sum_{i=1}^{n} \alpha_i$$
$$= -\frac{1}{2} \sum_{i=1}^{m} \sum_{j=1}^{m} \alpha_i \alpha_j d^i d^j \langle x^i, x^j \rangle + \sum_{i=1}^{n} \alpha_i.$$

Thus we arrive at the following, so-called *dual optimization problem* to the former *primal problem*:

MARGIN (dual version)
Maximize

$$-\frac{1}{2} \sum_{i=1}^{m} \sum_{j=1}^{m} \alpha_i \alpha_j d^i d^j \langle x^i, x^j \rangle + \sum_{i=1}^{m} \alpha_i$$

subject to $\alpha_1 \geq 0, \ldots, \alpha_m \geq 0$ and $\sum_{i=1}^{m} \alpha_i d^i = 0$.

As the function to be maximized is a quadratic, thus convex function, and constraints are linear, this is a simple standard task for numerical mathematics. What is remarkable is the fact that training vectors x^i always occur encapsulated within an inner product $\langle x^i, x^j \rangle$. This will soon turn out to be one of the main advantages of support vector machines.

7.2.5 Kernel Function Trick

Remember the starting point of this section, a training set of positive and negative points in n-dimensional real space for which we seek for a separating hyperplane. Most often, training sets are not linearly separable, but canonical embeddings in higher-dimensional spaces make them linearly separable. The simplest such problem is logical XOR. Drawing positive points $(1, 0)$ and $(0, 1)$, and negative points $(0, 0)$ and $(1, 1)$ into the plane, we immediately see that the problem is not linearly separable (Fig. 7.6 (a)).

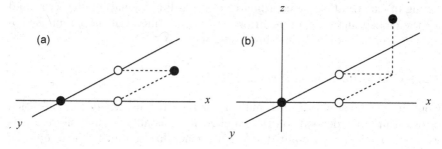

Fig. 7.6. XOR: **(a)** not linearly separable in 2 dimensions; **(b)** linearly separable after being suitably embedded into 3 dimensions

Embedding this 2-dimensional problem into 3-dimensional space by introducing as a third component $z = \text{AND}(x, y)$, leads to a linearly separable problem (Fig. 7.6 (b)) with positive points $(1, 0, 0)$ and $(0, 1, 0)$, and negative points $(0, 0, 0)$ and $(1, 1, 1)$. Embedding into higher dimensional spaces obviously improves the chance of obtaining a linearly separable problem, and it does not lead to efficiency problems in the dynamics of a support vector machine if one uses embeddings into higher dimensional spaces (even Hilbert spaces of infinite dimension) based on so-called *kernel functions*. By this it is meant that for an embedding $\Phi : \mathbb{R}^n \to \mathbb{R}^N$, inner products in N-dimensional space for embedded points $\langle \Phi(x), \Phi(y) \rangle$ can be computed from inner product in n-dimensional space by applying a kernel function k:

$$\langle \Phi(x), \Phi(y) \rangle = k(\langle x, y \rangle).$$

As a simple example, consider the embedding function Φ that maps an n-dimensional vector to all products of coordinates (this might make sense if correlations between coordinates play a role in a problem).

$$\Phi\left((x_i)_{i=1,\dots,n}\right) = (x_i x_j)_{i,j=1,\dots,n}$$

For $n = 2$ this looks as follows:

$$\Phi(x_1, x_2) = \left(x_1^2, x_1 x_2, x_2 x_1, x_2^2\right).$$

We reduce computation of inner products of embedded vectors in 4-dimensional space to inner products in 2-dimensional space as follows:

$$\langle \Phi(x_1, x_2), \Phi(y_1, y_2) \rangle = x_1^2 y_1^2 + x_1 x_2 y_1 y_2 + x_2 x_1 y_2 y_1 + x_2^2 y_2^2$$
$$= \langle (x_1, x_2), (y_1, y_2) \rangle^2$$

Here we used kernel function $k(z) = z^2$.

Exercise

7.2. Develop kernel functions for the embedding that forms all products consisting of exactly d of the coordinates of an n-dimensional input vector, and for the embedding that forms all products consisting of at most d of the coordinates of an n-dimensional input vector (for $d \le n$).

7.2.6 Generalization and Over-Fitting

Contrary to the situation with multi-layer perceptrons, where generalization error estimations depends on the number of adjustable parameters of a network (in particular on input dimension), generalization error for an optimal support vector machine trained on a set of m training points taken from a ball of radius R and having maximum margin μ can be shown to scale with

$$\frac{1}{m} \frac{R^2}{\mu^2}.$$

Note that the quotient $R\mu^{-1}$ must be considered, since scaling of training data does not alter the problem, of course, but leads also to a scaling of margin. Only by using quotient $R\mu^{-1}$ we are invariant with respect to scaling.

This result shows the remarkable fact that input dimension does not play any role with respect to generalization ability of a support vector machine. This allows us to introduce more liberally problem-specific input features. But note that introduction of additional features must not be done in a completely uncritical manner. Though maximum margin is expected to increase after introduction of additional features, also radius R of the ball where data are taken from will increase. Thus we must take care of a certain balance between increase of margin and of radius.

7.2.7 Application to Contact Map Prediction

The problem of contact map prediction for proteins was described in Sect. 2.10.1. Between predicting secondary structure of a protein P consisting of n amino acids $P = P(1) \ldots P(n)$ and predicting its precise 3D structure lies prediction of the so-called *contact map*. Contact map of a protein of n amino acids is a symmetric $n \times n$ matrix consisting of entries 0 and 1. If the entry in row i and column j is 1 this expresses that amino acids $P(i)$ and $P(j)$ are nearby up to a certain extend in the natural fold of the protein. We describe a support vector machine approach. As in Sect. 7.1, we focus on the design of the interface to a network. To predict an eventual contact between amino acids $P(i)$ and $P(j)$ of protein P, one again uses a certain number of neighbouring amino acids, for example the two neighbours to the left and the two neighbours to the right of each of $P(i)$ and $P(j)$. Besides certain static chemical information, for example polarities of $P(i)$ and $P(j)$, it is also convenient to use the above treated alpha-beta-coil prediction for each of these 10 considered amino acids. A further source of valuable information concerning contact behaviour within protein P comes from the analysis of mutational covariance at sites i and j within a multiple alignment of a family $\mathcal{P} = P_1, \ldots, P_m$ of homologous proteins to which protein P belongs (see Fig. 7.7). In case that $P(i)$ and $P(j)$ are in contact this usually means that $P(i)$ and $P(j)$ are part of a core structure of the protein that, being functionally relevant, is strongly conserved through all proteins of protein family $\mathcal{P}$. This should be observable by a positive covariance of mutational behaviour at sites i and j in the following sense.

Consider fixed sites i and j. We are interested in determining whether there is a significant covariance of mutational events at sites i and j through protein family $\mathcal{P}$. We assume that for any two pairs of amino acids A and B there is some numerical measure $d(A, B)$ that expresses how strongly A deviates from B. Considering at site i all pairs $P_p(i)$ and $P_q(i)$ of amino acids, for $p < q$, we compute the expected value of these pairwise distances at site i and j:

$$E_i[d] = \frac{2}{m(m-1)} \sum_{p<q} d\left(P_p(i), P_q(i)\right)$$

$$E_j[d] = \frac{2}{m(m-1)} \sum_{p<q} d\left(P_p(j), P_q(j)\right).$$

We now look whether being above-average or below-average for pairs of amino acids at site i co-occurs with being above-average or below-average for the corresponding pairs of amino acids at site j. Such a covariance is measured by the following sum:

$$\sum_{p<q} (d(P_p(i), P_q(i)) - E_i[d])(d(P_p(j), P_q(j)) - E_j[d]).$$

Usually this sum is normalized by division by variances at sites i and j

$$\sigma_i^2 = E_i[d(A,B)^2] - E_i[d(A,B)]^2$$
$$\sigma_j^2 = E_j[d(A,B)^2] - E_j[d(A,B)]^2$$

to finally obtain the following covariance coefficient:

$$\frac{1}{\sigma_i^2 \sigma_j^2} \sum_{p<q} (d(P_p(i), P_q(i)) - E_i[d(A,B)])(d(P_p(j), P_q(j)) - E_j[d(A,B)]).$$

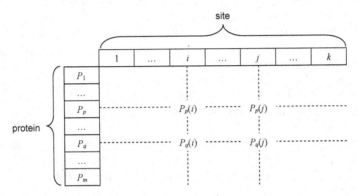

Fig. 7.7. Multiple alignment of a family of homologous proteins

Now we consider sites $i-2, i-1, i, i+1, i+2$ and $j-2, j-1, j, j+1, j+2$ and compute 25 pairwise covariance coefficients. These are presented to the support vector machine as additional features. This makes sense as, for example, for a parallel beta-sheet high positive values are expected in the main diagonal of the covariance matrix, and for an anti-parallel beta-sheet high positive values are expected in the main anti-diagonal of the covariance matrix (see Fig. 7.8).

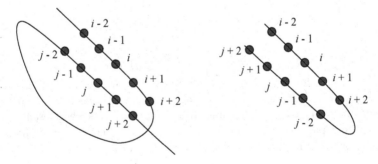

Fig. 7.8. Expected higher covariances between sites in parallel and anti-parallel beta-sheet

7.2.8 Soft Margin Maximization

In case of training points that are not linearly separable one attempts to separate them "as good as possible", with least classification error. This can be integrated in the formalism of quadratic minimization in a canonical manner as follows. Replace the strict constraints

$$d^i(\langle w, x^i \rangle + b) - 1 \geq 0$$

by the more liberal "soft constraints"

$$d^i(\langle w, x^i \rangle + b) - 1 + \eta_i \geq 0$$

with values $\eta_i \geq 0$ chosen as small as possible. It seems reasonable to minimize the sum of all extra values η_i, and also to control priority between minimizing $\|w\|^2$ and minimizing sum of extra values by a suitable selected constant C. Thus we arrive at the following generalized minimization problem:

SOFT MARGIN (1)
Given training set $T = (x^1, d^1), \ldots, (x^m, d^m)$ minimize (over w, b, η, with suitably chosen constant C)

$$\frac{1}{2}\|w\|^2 + C \sum_{i=1}^{m} \eta_i$$

subject to $\eta_i \geq 0$ and $d^i(\langle w, x^i \rangle + b) - 1 + \eta_i \geq 0$.

This can be transformed into dual form such that again the "kernel trick" gets applicable. Replacing quadratic vector norm $\|w\|^2$ by linear norm $\|w\|_1 = \sum_{k=1}^{n} |w_k|$ allows a further simplification of the problem towards linearity as a so-called "linear program":

SOFT MARGIN (2)
Given training set $T = (x^1, d^1), \ldots, (x^m, d^m)$, minimize (over w, b, η, β, with suitably chosen constant C)

$$\frac{1}{2} \sum_{k=1}^{n} \beta_k + C \sum_{i=1}^{m} \eta_i$$

subject to $\eta_i \geq 0$, $\beta_k \geq 0$, $-\beta_k \leq w_k$, $w_k \leq \beta_k$ and $d^i(\langle w, x^i \rangle + b) - 1 + \eta_i \geq 0$.

7.2.9 Semi-Supervised Support Vector Machines

In bioinformatics classification problems there are often labelled data, as well as a couple of unlabelled data. Stated in terms of statistical learning theory, we have the situation where the distribution of data is not completely unknown, but to some extend a-priori known. In such a situation one is tempted to try to incorporate such knowledge into training algorithms. In case of support vector machines, this is easily achievable (we refer to [10, 18, 17]). We make the assumption that unseen data will probably group more or less close around labelled data such that is makes sense to look for an assignment of labels to unseen data in a such a way that still a margin as large as possible is achievable with a support vector machine. Counting point y^j as a positive point, some minimal value $\xi_j \geq 0$ should assist $(\langle w, y^j \rangle + b) - 1$ to get non-negative, that is $(\langle w, y^j \rangle + b) - 1 + \xi_j \geq 0$, whereas counting point y^j as a negative point, some minimal value $\zeta_j \geq 0$ should assist $-(\langle w, y^j \rangle + b) - 1$ to get positive, that is $-(\langle w, y^j \rangle + b) - 1 + \zeta_j \geq 0$. We should prefer the counting of y^j as a positive or negative point that is cheaper, measured by $\min\{\xi_j, \zeta_j\}$. In summary, we end with the following minimization problem:

SEMI-SUPERVISED SOFT MARGIN (1)

Given labelled training data $(x^1, d^1), \ldots, (x^m, d^m)$ and unlabelled training data $y^1, \ldots, y^M$, minimize (over w, b, η, ξ, ζ, with suitably chosen constant C)

$$\frac{1}{2}\|w\|^2 + C \left(\sum_{i=1}^{m} \eta_i + \sum_{j=1}^{M} \min\{\xi_j, \zeta_j\} \right)$$

subject to $\eta_i \geq 0$, $\xi_j \geq 0$, $\zeta_j \geq 0$ and

$$d^i(\langle w, x^i \rangle + b) - 1 + \eta_i \geq 0,$$

$$\langle w, x^j \rangle + b - 1 + \xi_j \geq 0,$$

$$-(\langle w, x^j \rangle + b) - 1 + \zeta_j \geq 0.$$

The unusual min-term may be replaced by a linear term by using a binary decision variable b^j that expresses whether point y^j is counted positive or negative.

SEMI-SUPERVISED SOFT MARGIN (2)

Given labelled training data $(x^1, d^1), \ldots, (x^m, d^m)$ and unlabelled training data $y^1, \ldots, y^M$, minimize (over w, b, η, d, with suitably chosen constants C and D)

$$\frac{1}{2}\|w\|^2 + C\left(\sum_{i=1}^{m} \eta_i + \sum_{j=1}^{M}(\xi_j + \zeta_j)\right)$$

subject to $\eta_i \geq 0$, $\xi_j \geq 0$, $\zeta_j \geq 0$, $d^j \in \{0,1\}$ and

$$d^i(\langle w, x^i \rangle + b) - 1 + \eta_i \geq 0,$$

$$\langle w, x^j \rangle + b - 1 + \xi_j + D(1 - d^j) \geq 0,$$

$$-(\langle w, x^j \rangle + b) - 1 + \zeta_j + Dd^j \geq 0.$$

Look what happens for the case $d^j = 1$. In this case, y^j is counted as a positive point. Constraint $\langle w, x^j \rangle + b - 1 + \xi_j + D(1 - d^j) \geq 0$ reduces to the former one that is required for any positive point, $\langle w, x^j \rangle + b - 1 + \xi_j \geq 0$. Choosing D sufficiently large guarantees that the second constraint $-(\langle w, x^j \rangle + b) - 1 + \zeta_j + Dd^j \geq 0$ does not lead to a further extra term $\zeta_j \geq 0$ by being satisfiable already with the least possible value $\zeta_j = 0$. Similarly, look what happens for the case $d^j = 0$. In this case, y^j is counted as a negative point. Now constraint $-(\langle w, x^j \rangle + b) - 1 + \zeta_j + Dd^j \geq 0$ reduces to the former one that is required for any negative point, $\langle w, x^j \rangle + b - 1 + \xi_j \geq 0$. Choosing D sufficiently large guarantees that the first constraint $\langle w, x^j \rangle + b - 1 + \xi_j + D(1 - d^j) \geq 0$ does not lead to a further extra term $\xi_j \geq 0$ by being satisfiable already with the least possible value $\xi_j = 0$.

However, note that we did not end with a linear program (that would allow an efficient solution), but with an integer program due to the fact that variable d^j takes binary values. Nevertheless, one might apply heuristics or approximation algorithms known to be applicable to integer programming problems. In that sense, the problem formulation obtained above as an integer program is preferable over a plain full search through all 2^M possibilities to partition the unlabelled points into two classes, positive and negative.

7.3 Hidden Markov Models

Hidden Markov models have been already applied in former chapters. We introduce here the formal details of what Hidden Markov models are, and discuss further applications.

7.3.1 Architecture of Hidden Markov Models

One best thinks of Hidden Markov models as directed graphs whose nodes represent *hidden states* with a probability distribution of emission probabilities over characters x of a finite alphabet attached to each state, and directed edges between nodes labelled with non-zero *transition probabilities*. Denote the transition probability from state p to state q by T_{pq}, and the emission probability for character x in state q by $E_q(x)$. There is a distinguished *start node* $q(0)$ (and sometimes also a distinguished *final node*).

7.3.2 Causes and Effects

The goal of a Hidden Markov model is to probabilistically generate strings of emitted characters by starting at node $q(0)$, and then walking through nodes (hidden states) according to the transition probabilities. Separating between the inner world of hidden states and outer world of observed characters is the main source of flexibility that Hidden Markov models offer to a user. Besides this, the option of modelling local interdependencies between adjacent positions in an emitted string is the second source of model flexibility. But note that the option to model long range interdependencies is not available in Hidden Markov models. There are five basic probability distributions that play a central role in Hidden Markov model usage. Calling hidden state sequences "causes" and emitted character sequences "effects", these distributions aim to measure with which probability

- a cause occurs together with an effect (*joint probability*)
- a given cause produces effects (*conditional probability of effects, given a cause*)
- a cause occurs (*marginal probability*)
- an effect occurs (*marginal probability*)
- a cause occurs, given an effect (*conditional probability of causes, given an effect*)

Formal notation and definitions are as follows, for any state sequence $Q = q(0)q(1)\ldots q(n)$ starting with initial state $q(0)$ and character sequence $S = x_1\ldots x_n$ (we assume, for simplicity, that all occurring values in denominators are greater than zero).

$$P(Q) = \prod_{i=0}^{n-1} T_{q(i)q(i+1)}$$

marginal distribution of causes

$$P(S \mid Q) = \prod_{i=1}^{n} E_{q(i)}(x_i)$$

conditional distribution of effects, given cause

$$P(Q,S) = P(Q)P(S|Q)$$

joint distribution

$$P(S) = \sum_Q P(S,Q)$$

marginal distribution of effects

$$P(Q \mid S) = \frac{P(Q,S)}{P(S)}$$

conditional distribution of causes, given effect

The conditional distribution of effects, given causes, describes the model. It can be estimated by simply let the model run and sample data. The marginal distribution describes a priori, that is before an observation is made, knowledge about the occurrence of causes. The conditional distribution of causes, given effects, describes a posteriori, that is after having made an observation, knowledge about the presence of a certain cause. The latter is what one is interested in. For example, observing strings, a most probable state sequence in the Hidden Markov model discussed in Sect. 3.7 might define a plausible multiple alignment of strings, or observing an DNA string, a most probable state sequence in a properly designed and trained Hidden Markov model might define which parts of the string are exons of a gene.

Conditional probabilities are related by Bayes rule:

$$P(Q \mid S) = \frac{P(S \mid Q)P(Q)}{P(S)}.$$

Given observation sequence S, the Viterbi algorithm allows efficient computation of a most probable state sequence Q, that is, state sequence Q such that the following holds:

$$P(Q \mid S) = \max_{Q'} P(Q' \mid S).$$

Whereas the definitions of the probabilities above immediately can be used for an efficient computation for the cases of $P(Q,S)$, $P(S \mid Q)$, and $P(Q)$, the definition of $P(S)$, and thus also the definition of $P(Q \mid S)$, involve a sum of exponential many terms. Using the forward variable introduced in Sect. 3.8 allows replacement of this inefficient computation by an efficient one:

$$P(S) = \sum_{q \in Q} \alpha_n(q).$$

Table 7.1. CG island: transition probabilities inside and outside

+	A	C	G	T	-	A	C	G	T
A	0.180	0.274	0.426	0.120	A	0.300	0.205	0.285	0.210
C	0.171	0.368	0.274	0.187	C	0.322	0.298	0.078	0.302
G	0.161	0.339	0.375	0.125	G	0.248	0.246	0.298	0.208
T	0.079	0.355	0.384	0.182	T	0.177	0.239	0.292	0.292

7.3.3 Bioinformatics Application: CG Islands

In genomes, substring CG occurs more frequently in promoter or start regions of genes than in non-coding regions. Regions with a high proportion of CG occurrences are called *CG islands*. Table 7.1 shows for each base X the probability that it is followed by base Y in CG islands (+) and non CG islands (-).

Besides these conditional probabilities that we denote with $P(Y|X, +)$ and $P(Y|X, -)$ we also use information about the probabilites of occurrence of a single base X in CG islands and non CG islands and denote these by $P(X|+)$ and $P(Y|-)$. The difference in the frequency of occurrence of CG in CG islands and non CG islands can be used to predict genes. Simply let a window of length k, for a certain number k, walk over a given DNA string and for any such window string S of length k compute the probability $P(S|+)$ that it is generated by the CG island process, and compare it to the corresponding probability for the non CG island process. These probabilities, which are also called likelihood of the corresponding models given S, are calculated by

$$L(+|S) = P(S|+) = P(S(1)|+) \prod_{i=1}^{k-1} P(S(i+1)|S(i), +)$$

$$L(-|S) = P(S|-) = P(S(1)|-) \prod_{i=1}^{k-1} P(S(i+1)|S(i), -).$$

The model with higher likelihood is chosen. There are two problems with this attempt to predict CG islands. First, there must be a choice of window length k. It is not clear how to best choose k. Furthermore, a variable window length could be better than a fixed length over the whole genome string. Second, when letting the window walk along the genome string, likelihood of CG island or non CG island may change gradually. We are left with the decision where to set start and end points for expected genes. Both problems do not occur when using Hidden Markov model prediction instead of likelihood comparison. In our example, a suitable Hidden Markov model has internal states that are denoted $X(+)$ and $X(-)$, with X ranging over A, C, G, T. There are probabilities for an initial occurrence of any of these states, as well as transition probabilities between states. Finally there are symbols that are emitted by the states with certain probabilities. In our example, the emission probabilities are very special in the sense that $X(+)$ emits symbol X with certainty, as does $X(-)$. In general, emission of symbols is more complex than in this example. Note that internal states are not visible to the user, whereas emitted symbols can be observed. In our example, state $X(+)$ represents the generation of symbol X within a CG island, and state $X(-)$ the generation of symbol X in a non CG island.

7.3.4 Further Applications

As was already the case for neural networks, the most comprehensive overview of Hidden Markov model applications to bioinformatics is again Baldi & Brunak [4]. Applications deal with the identification and prediction of, for example:

- Signal peptides and signal anchors [58]
- Promoter regions [62]
- Periodic patterns in exons and introns [6]
- G-protein-coupled receptors [7]
- Genes in E.coli [48]
- Human genes [5]
- Protein modelling [47]

7.4 Genetic Algorithms

Genetic algorithms attempt to solve optimization problems by mimicking the process of adapting a population of organisms to its environment by the operations of mutation, recombination of genetic material (DNA), and selection of fittest members of a population.

7.4.1 Basic Characteristics of Genetic Algorithms

The first step in solving an optimization problem by a genetic algorithm is to encode problem solutions as bit strings (sometimes also strings over more extended alphabets are used). Whereas *bit string encoding* of solutions usually is a rather canonical process, sometimes it may be more sophisticated. Some care has to be taken to ensure that mutations and recombinations always lead to bit strings that again encode admissible solutions. Next step is to define fitness of bit strings. Here is the point where the function to be optimized in an optimization problem finds its expression. Having done this, the algorithm is rather simple. An *initial population of strings* is generated. Its size is variable and depends on the problem instance. It serves as parental population for the next step of the algorithm. Typically, the initial population is generated randomly, but it is also possible to place bit strings of the initial population in areas where optimal solutions are expected. After having generated the parental population one ranks individuals according to their *fitness* and selects those that are allowed to propagate their characteristics to the next generation. The fitter an individual is, the bigger are its chances to give the genetic material to its descendants. The expectation is that individuals who are closer to an optimal solution will finally be preferred over those that are less optimal. Most selection functions are stochastic so that also a small proportion of less optimal individuals are selected. This keeps diversity of the population large and prevents early convergence. It guarantees that the algorithm does not get stuck in local optima at an early stage. The simplest and most often used selection function is *roulette wheel selection*: if f_i is the fitness of individual i, its probability of being selected is

$$p_i = \frac{f_i}{\sum_{j=1}^{N} f_j}$$

where N is the number of individuals in the population. For producing new solutions, pairs of selected solutions are modified by the processes of *recombination* and *mutation*. This leads to a propagation of characteristics from parents to their descendants, but also to the formation of new characteristics. Recombination is the counterpart of sexual reproduction: in nature, chromosomes pair up during meiosis and exchange parts of its genetic material; in a genetic algorithm, new solutions are formed by cutting two selected strings at the same randomly chosen position and exchanging suffixes, a process called *one-point crossover* (see Fig. 7.9). Often, genetic algorithms use variants of this process, like two-point crossover or "cut and splice".

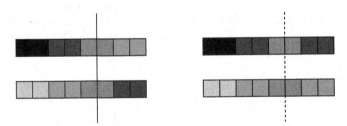

Fig. 7.9. The scheme of one-point crossover

The process of *mutation* is analogous to biological mutation. It is the source for new genetic material. In biological evolution there are different kinds of mutations, but genetic algorithm only use single-bit substitution as shown in Fig. 7.10.

Whereas the processes of recombination and mutation are strongly affected by chance, the selection of the next population gives the genetic algorithm a direction towards optimality of solutions. Recombination/mutation loop is repeated until some termination condition is satisfied. Some possible termination conditions are:

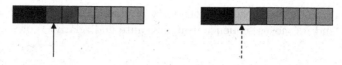

Fig. 7.10. The scheme of one-point mutation

- A solution which is close enough to the optimum is found.
- A fixed number of generations has been computed.
- There is no progress within a fixed number of generations.
- A manual check shows good solutions.

7.4.2 Application to HP-Fold Optimization

The problem of protein structure prediction was described in Sect. 1.7. A common approach uses the HP-model. Amino acids are divided into hydrophobic (H) and polar (P) amino acids. In the process of forming a tertiary structure, the hydrophobic amino acids tend to form non-covalent bonds since these lead to stable conformations of low free energy. The HP-model is a lattice model. The amino acids of a protein are placed on the vertices of a grid such that consecutive amino acids are placed on the grid side by side. Figure 7.11 shows several grids the model can work on. A non-covalent binding occurs between two hydrophobic amino acids in case that they are not consecutive in the protein and are placed side by side on the grid.

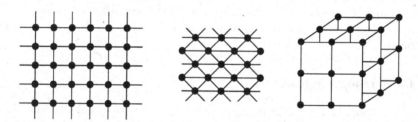

Fig. 7.11. The structure of three different grids the HP-model can use is shown. Starting on the left hand side we see a quadratic, a triangular, and a cubic conformation.

A genetic algorithm may be used to find favourable HP-model conformations. In the following presentation a 2-dimensional rectangular grid is used. A standard way to represent a protein conformation on the grid is the use of a string of *moves* on the grid, with moves being represented in an obvious manner as directions 'U' (up), 'D' (down), 'L' (left), and 'R' (right). Besides this "absolute representation" of moves, alternatively a "relative representation" of moves may be used: here 'U' means to move the same direction as in the step before, 'L' means to change formerly used direction to the left, and 'R' to change formerly used direction to the right. Figure 7.12 shows a HP-sequence on a grid and its moves sequence using absolute and relative representation.

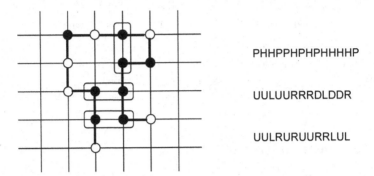

PHHPPHPHPHHHHP

UULUURRRDLDDR

UULRURUURRLUL

Fig. 7.12. A protein conformation on a lattice model is displayed. On the right side the HP-sequence, the absolute moving step sequence, and the relative moving step for this protein is shown.

Depending on the choice of absolute or relative encoding the operations of the genetic algorithm are different. For example, a one-point mutation under absolute encoding of move sequences requires a two-point mutation under relative encoding of move sequences to achieve the same effect. Next we have to define a fitness function on protein conformations. Following the discussion above on the effect of H-H pairs on free energy minimization we simply count the number of adjacent, but non-consecutive H-H pairs. After we have encoded the HP-model this way, a genetic algorithm as described in Sect. 7.4.1 can be applied.

7.4.3 Further Applications

Genetic algorithms became popular in the early 1970s thanks to the work of John Holland [36]. They have been successfully applied to solve problems from lots of fields. Applications in bioinformatics cover, for example:

- Multiple sequence alignment [30, 59]
- Building phylogenetic trees [66]
- Molecular docking [80]

7.5 Ant Colony Optimization

Like genetic algorithms, ant algorithms are heuristic search algorithms. They mimic the behaviour of ants. In ant colonies one observes that ants rapidly find the shortest path between the colony and some food source. Whereas a single ant would not be able to do so, the cooperation of the whole colony of ants solves the task. This is an example of *swarm intelligence*. The Ant Colony Optimization algorithm (ACO) is the most famous among such algorithms.

7.5.1 Basic Characteristics of Ant Colony Algorithms

Whenever several individuals work together there has to be some form of communication. Communication among ants is done by placing scents called pheromones on trails that ants use. Whenever an ant finds a food source it walks back to the colony and alarms the other ants. These then follow the pheromone marked trail towards the food source and, by this way, strengthen the trail by putting further pheromone onto it. By this simple process it is possible that the whole colony helps to exploit the found food source. Look what happens if a trail is destroyed by placing a barrier on it. Now the ants are forced to find a new path to the food source, and they usually find the shortest one. First, they try to find a way around the barrier without any orientation until a few of them recover the pheromone trail onto the other side of the barrier, thus following the existing scents to the food source. As soon as the first ant comes back, lots of pheromone trails leading back to the colony have been randomly generated. While some ants found their way to the food source and walked back probably more ants have passed the shorter path around the barrier than the longer one. So the ants returning from the food source prefer the shorter path back to the colony. Again, that increases the amount of pheromone placed on the shorter trail, and more and more ants are going to choose that shorter path. This process is illustrated in Fig. 7.13. As pheromone disappears the longer one of the trails gets more and more unattractive. This also happens when the food source runs dry.

This episode shows that a single ant may well find the food source and back to the colony. But only a huge number of ants is able to find, with high probability, the shortest path between some food source and the colony. Communication among the individuals of a colony proceeds by modifications of the environment: marking trails by pheromone in natural ant colonies, and modifying certain globally available numeric parameters in artificial ant colony algorithms. Pheromone evaporation is responsible for the deletion of suboptimal solutions in the course of time. Unlike ants in nature, artificial ants can have additional characteristics (simplifying the handling of the ACO algorithm), as for example:

- A memory for storing paths
- Different pheromones depending on the quality of the found solution
- Additional functions like local optimization or backtracking

Formally, artificial ACO is a search technique working on a graph structure. It starts with a colony of artificial ants. Each ant follows a trail (generates a solution) depending on global information and random processes. Furthermore, it modifies global memory depending on the quality of the found solution.

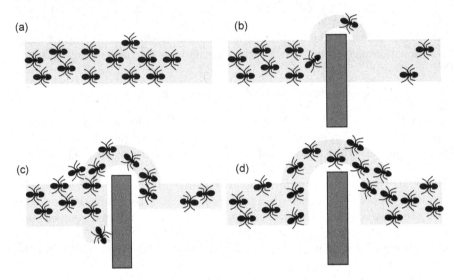

Fig. 7.13. The behavior of ants is shown. **(a)** Ants form trails to exploit food sources using scents to mark the way. **(b)** When a barrier is placed on their trail they are unoriented. **(c)** The ants start trying to round the balk until they recover the pheromone trail. In a fixed time slice more ants make their way on the shorter trail than on the longer one. So more pheromones will be set there. **(d)** Because ants prefer the trail with the most pheromones, the shorter trail will be preferred more and more.

Other ants following explored trails include this global information in their selection of alternative trails and modifications of the environment.

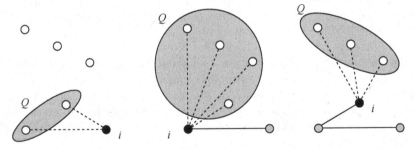

Fig. 7.14. Computation of a solution is shown step by step. Starting at vertex i, the probability for choosing the next vertex out of all relevant vertices $q \in Q$ will be computed. By means of this probability the following vertex is chosen. Thereafter, the procedure is repeated with the chosen vertex as the new start point i until the solution is complete.

The probability for walking from actual vertex i to next vertex j is given by some value p_{ij}. It depends on two parameters. The first one is the called *attractiveness* η_{ij} of the step; it represents some a priori preference for selecting the step. This parameter can be appreciated as environmental impact on the behaviour of the ants. The second one is the *trail level* τ_{ij}. It represents the a posteriori preference of the step expressing how profitable the step from vertex i to vertex j was in the past. This parameter corresponds to the pheromone rate on the trail. Probability p_{ij} is then defined as follows:

$$p_{ij} = \frac{\tau_{ij}^{\alpha}\eta_{ij}^{\beta}}{\sum_{q \in Q}\left(\tau_{iq}^{\alpha}\eta_{iq}^{\beta}\right)}.$$

Here, exponent parameters α and β control how strong a priori and a posteriori preferences of steps influence the decision. Q denotes the set of all relevant vertices that are accessible from vertex i. After all ants have, based on these probablitites, generated their solutions the trail levels will be updated as follows:

$$\tau_{ij} = \rho\tau_{ij} + \Delta\tau_{ij}.$$

Here, ρ is a user-defined coefficient that represents evaporation of pheromone, whereas $\Delta\tau_{ij}$ expresses how strongly preference of the move from vertex i to vertex j increases. Computing solutions and updating trail levels proceeds until a termination condition is satisfied. Some possible termination conditions are:

- Solution which is close enough to the optimum is found
- Fixed number of iterations is computed
- There is no progress during a fixed number of iterations
- Manual check shows good solutions

7.5.2 Application to HP-Fold Optimization

Protein folding is mainly the process of taking a conformation of minimum free energy. As an algorithmic problem, protein tertiary structure prediction is NP-hard. The HP-model established in Sect. 1.7 and described in detail in Sect. 7.4.2 is one method for minimizing the free energy. The amino acids are divided into hydrophobic (H) and polar (P) amino acids. Tertiary structure formation is mainly driven by the forming of non-covalent bindings between hydrophobic amino acids. This kind of binding is energetically favourable. Thus we are looking for a conformation with as many as possible non-covalent bindings between hydrophobic amino acids. The Ant Colony Optimization algorithm (ACO) can be used to solve the problem of protein tertiary structure prediction. We describe an ACO working on the 2-dimensional grid of the HP-model. The artificial ants construct possible folds of a given sequence. Each ant starts the folding process at a randomly chosen start point

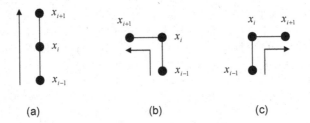

Fig. 7.15. The possible conformations of an amino acid triple (x_{i-1}, x_i, x_{i+1}) are shown. You can see the three (relative) move directions **(a)** 'U' (up), **(b)** 'L' (left) and **(c)** 'R' (right).

within the protein. A fold is then extended into both directions of the protein. Actually sitting at a grid position i, the extension of the conformation to grid position j is done with probability that is dependent on the attractiveness $\eta_{i,j}$ and the trail level $\tau_{i,j}$ of the move. The heuristic function $\eta_{i,j}$ guarantees that high quality solutions are preferred. The attractiveness $\eta_{i,j}$ depends on the number of non-covalent bindings between two hydrophobic amino acids that are generated by the move. The more non-covalent bindings occur by moving to j the more attractive will this move be. A difficulty of this simple approach is that some agents may not find solutions because they achieved a position in the grid where all neighbouring vertices are already occupied. To deal with this may be done by a backtracking mechanism.

7.5.3 Further Applications

Ant Colony Optimization was introduced in 1992 by Marco Dorigo. One of the benchmarks of Ant Colony Optimization is described in [26]. Like genetic algorithms, ACO's found application in various fields. Applications in bioinformatics cover, for example:

- Multiple sequence alignment [19]
- Protein/ligand docking [45]

7.6 Bibliographic Remarks

Among lots of good books introducing the theory of neural networks we mention only one, Bishop & Hinton [12]. The classical introduction to support vector machines is Cristianini & Shawe-Taylor [24]. The most comprehensive textbook on machine learning approaches with an emphasis on neural networks and Hidden Markov models is clearly Baldi & Brunak [4]. An extended introduction to formal foundations of statistical learning theory can be found in Vapnik [73] and Vidyasagar [76], as well as Cristianini & Shawe-Taylor [24]. Focus on Hidden Markov models applied to bioinformatics is laid in Koski [46]. A good introduction to ant algorithms is Dorigo et al. [25], a classical textbook on genetic algorithms is Mitchell [54].

References

[1] Eric L. Anson and Eugene W. Myers. Algorithms for whole genome shotgun sequencing. In *RECOMB*, pages 1–9, 1999.

[2] Sanjeev Arora and Shmuel Safra. Probabilistic Checking of Proofs: A New Characterization of NP. In *33rd Annual Symposium on Foundations of Computer Science*, pages 2–12. IEEE, 1992.

[3] Vineet Bafna, Eugene L. Lawler, and Pavel A. Pevzner. Approximation Algorithms for Multiple Sequence Alignment. *Theor. Comput. Sci.*, 182(1-2):233–244, 1997.

[4] Pierre Baldi and Søren Brunak. *Bioinformatics: The Machine Learning Approach (Adaptive Computation and Machine Learning)*. The MIT Press, 1998.

[5] Pierre Baldi, Søren Brunak, Yves Chauvin, Jacob Engelbrecht, and Anders Krogh. Hidden markov models for human genes. In *NIPS*, pages 761–768, 1993.

[6] Pierre Baldi, Søren Brunak, Yves Chauvin, Jacob Engelbrecht, and Anders Krogh. Periodic sequence patterns in human exons. In *ISMB*, pages 30–38, 1995.

[7] Pierre Baldi and Yves Chauvin. Hidden Markov Models of the G-Protein-Coupled Receptor Family. *Journal of Computational Biology*, 1(4):311–336, 1994.

[8] Richard Bellman. *Dynamic Programming*. Princeton University Press, 1957.

[9] Richard Bellman. *Dynamic Programming*. Dover Publications Inc., 2003.

[10] Kristin P. Bennett and Ayhan Demiriz. Semi-supervised support vector machines. In *NIPS*, pages 368–374, 1998.

[11] Bonnie Berger and Frank Thomson Leighton. Protein folding in the hydrophobic-hydrophilic (*P*) is NP-complete. In *RECOMB*, pages 30–39, 1998.

[12] C.M. Bishop and Geoffrey Hinton. *Neural Networks for Pattern Recognition*. Clarendon Press, 1995.

[13] Kellogg S. Booth and George S. Lueker. Testing for the Consecutive Ones Property, Interval Graphs, and Graph Planarity Using PQ-Tree Algorithms. *J. Comput. Syst. Sci.*, 13(3):335–379, 1976.

[14] Søren Brunak, Jacob Engelbrecht, and Steen Knudsen. Prediction of human mRNA donor and acceptor sites from the DNA sequence. *Journal of Computational Biology*, 220:49–65, 1991.

[15] Alberto Caprara. Sorting by reversals is difficult. In *RECOMB*, pages 75–83, 1997.

[16] Humberto Carillo and David Lipman. The multiple sequence alignment problem in biology. *SIAM J. Appl. Math.*, 48(5):1073–1082, 1988.

[17] Oliver Chapelle, Bernhard. Schölkopf, and Alexander Zien. *Semi-Supervised Learning*. MIT Press, Cambridge, MA, 2006.

[18] Olivier Chapelle, Vikas Sindhwani, and S. Sathiya Keerthi. Branch and bound for semi-supervised support vector machines. In *NIPS*, pages 217–224, 2006.

[19] Ling Chen, Lingjun Zou, and Juan Chen. An efficient algorithm for multiple sequences alignment based on ant colony algorithms. In Hamid R. Arabnia, Jack Y. Yang, and Mary Qu Yang, editors, *GEM*, pages 191–196. CSREA Press, 2007.

[20] Peter Clote and Rolf Backofen. *Computational Molecular Biology: An Introduction (Wiley Series in Mathematical & Computational Biology)*. John Wiley & Sons, 2000.

[21] International Human Genome Sequencing Consortium. Initial sequencing and analysis of the human genome. *Nature*, 42:860–921, 2001.

[22] Stephen A. Cook. The Complexity of Theorem-Proving Procedures. In *STOC*, pages 151–158, 1971.

[23] Pierluigi Crescenzi, Deborah Goldman, Christos H. Papadimitriou, Antonio Piccolboni, and Mihalis Yannakakis. On the Complexity of Protein Folding. *Journal of Computational Biology*, 5(3):423–466, 1998.

[24] Nello Cristianini and John Shawe-Taylor. *An Introduction to Support Vector Machines and Other Kernel-based Learning Methods*. Cambridge University Press, 2000.

[25] Marco Dorigo, Gianni Di Caro, and Luca Maria Gambardella. Ant Algorithms for Discrete Optimization. *Artificial Life*, 5(2):137–172, 1999.

[26] Marco Dorigo and Thomas Stützle. *Ant Colony Optimization*. MIT Press, Cambridge, MA, 2004.

[27] Richard Durbin, Sean R. Eddy, Anders Krogh, and Graeme J. Mitchison. *Biological Sequence Analysis: Probabilistic Models of Proteins and Nucleic Acids*. Cambridge University Press, 1998.

[28] John Gallant, David Maier, and James A. Storer. On Finding Minimal Length Superstrings. *J. Comput. Syst. Sci.*, 20(1):50–58, 1980.

[29] Michael R. Garey and David S. Johnson. *Computers and Intractability: A Guide to the Theory of NP-completeness (Series of Books in the Mathematical Sciences)*. W.H.Freeman & Co Ltd, 1979.

[30] C. Gondro and B. P. Kinghorn. A simple genetic algorithm for multiple sequence alignment. *Genet. Mol. Res.*, 6(4):964–982, 2007.

[31] Dan Gusfield. *Algorithms on Strings, Trees and Sequences: Computer Science and Computational Biology.* Cambridge University Press, 1997.

[32] Sridhar Hannenhalli and Pavel A. Pevzner. Transforming cabbage into turnip: polynomial algorithm for sorting signed permutations by reversals. In *STOC*, pages 178–189, 1995.

[33] Sridhar Hannenhalli and Pavel A. Pevzner. Transforming men into mice (polynomial algorithm for genomic distance problem). In *FOCS*, pages 581–592, 1995.

[34] Dov Harel and Robert Endre Tarjan. Fast Algorithms for Finding Nearest Common Ancestors. *SIAM J. Comput.*, 13(2):338–355, 1984.

[35] William E. Hart and Sorin Istrail. Fast Protein Folding in the Hydrophobic-Hydrophillic Model within Three-Eights of Optimal. *Journal of Computational Biology*, 3(1):53–96, 1996.

[36] John H. Holland. *Adaptation in natural and artificial systems.* MIT Press, Cambridge, MA, USA, 1992.

[37] Thomas J. Hudson and et al.An STS-based map of the humane genome. *Science*, 270:1945–1954, 1995.

[38] Richard M. Karp. Reducibility among combinatorial problems. In *Complexity of Computer Computations*, pages 85–103. 1972.

[39] John D. Kececioglu and Eugene W. Myers. Combinatiorial Algorithms for DNA Sequence Assembly. *Algorithmica*, 13(1/2):7–51, 1995.

[40] John D. Kececioglu and R. Ravi. Of mice and men: Algorithms for evolutionary distances between genomes with translocation. In *SODA*, pages 604–613, 1995.

[41] John D. Kececioglu and David Sankoff. Exact and approximation algorithms for the inversion distance between two chromosomes. In *CPM*, pages 87–105, 1993.

[42] John D. Kececioglu and David Sankoff. Efficient bounds for oriented chromosome inversion distance. In *CPM*, pages 307–325, 1994.

[43] John D. Kececioglu and David Sankoff. Exact and Approximation Algorithms for Sorting by Reversals, with Application to Genome Rearrangement. *Algorithmica*, 13(1/2):180–210, 1995.

[44] Donald E. Knuth, James H. Morris Jr., and Vaughan R. Pratt. Fast Pattern Matching in Strings. *SIAM J. Comput.*, 6(2):323–350, 1977.

[45] Oliver Korb, Thomas Stützle, and Thomas E. Exner. An ant colony optimization approach to flexible protein–ligand docking. *Swarm Intelligence*, 1(2):115–134, 2007.

[46] Timo Koski. *Hidden Markov Models for Bioinformatics (Computational Biology).* Kluwer Academic Publishers, 2001.

[47] Anders Krogh, Michael Brown, I. Saira Mian, K. Sjölander, and David Haussler. Hidden Markov Models in Computational Biology: Applications to Protein Modelling. *Journal of Molecular Biology*, 235:1501–1531, 1994.

[48] Anders Krogh, I. Saira Mian, and David Haussler. Hidden Markov Models that find Genes in E.Coli DNA. *Nucleic Acids Research*, 22:4768–4778, 1994.

[49] Anders Krogh and Søren Kamaric Riis. Prediction of beta sheets in proteins. In *NIPS*, pages 917–923, 1995.

[50] Richard E. Lathrop. The protein threading problem with sequence amino interaction preferences is NP-complete. *Protein Engineering*, 7(9):1059–1068, 1994.

[51] M. Li, B. Ma, and L. Wang. On the Closest String and Substring Problems. *J. ACM*, 49(2):157–171, 2002.

[52] R.B. Lyngso and C.N. Pedersen. RNA pseudoknot prediction in energy based models. *J. Comput. Biol.*, 7(3-4):409–427, 2001.

[53] Martin Middendorf. More on the Complexity of Common Superstring and Supersequence Problems. *Theor. Comput. Sci.*, 125(2):205–228, 1994.

[54] Melanie Mitchell. *An Introduction to Genetic Algorithms (Complex Adaptive Systems)*. MIT Press, 1998.

[55] Edward M. McCreight. A space-economical suffix tree construction algorithm. *J. ACM*, 23(2):262–272, 1976.

[56] S. B. Needleman and C. D. Wunsch. A general method applicable to the search for similarities in the amino acid sequence of two proteins. *Journal of Molecular Biology*, 48:443–453, 1970.

[57] Henrik Nielsen, Jacob Engelbrecht, Søren Brunak, and Gunnar von Heijne. A Neural Network Method for Identification of Prokaryotic and Eukaryotic Signal Peptides and Prediction of their Cleavage Sites. *Int. J. Neural Syst.*, 8(5-6):581–599, 1997.

[58] Henrik Nielsen and Anders Krogh. Prediction of signal peptides and signal anchors by a hidden markov model. In *ISMB*, pages 122–130, 1998.

[59] C. Notredame and D. G. Higgins. SAGA: sequence alignment by genetic algorithm. *Nucleic Acids Res.*, 24(8):1515–1524, 1996.

[60] Ruth Nussinov and Ann B. Jacobson. Fast algorithm for predicting the secondary structure of single stranded RNA. *Proceedings of the National Academy of Sciences USA*, 71(11):6309–6313, 1980.

[61] Christos H. Papadimitriou. *Computational Complexity*. Addison Wesley, 1994.

[62] Anders Gorm Pedersen, Pierre Baldi, Søren Brunak, and Yves Chauvin. Characterization of prokaryotic and eukaryotic promoters using hidden markov models. In *ISMB*, pages 182–191, 1996.

[63] René Peeters. The maximum edge biclique problem is NP-complete. *Discrete Applied Mathematics*, 131(3):651–654, 2003.

[64] Pavel A. Pevzner. *Computational Molecular Biology: An Algorithmic Approach*. The MIT Press, 2000.

[65] N. Qjan and T. J. Sejnowski Predicting the secondary structure of globular proteins using neural network models. *Journal of Molecular Biology*, 202:865–884, 1988.

[66] T. H. Reijmers, Ron Wehrens, and Lutgarde M. C. Buydens. Quality criteria of genetic algorithms for construction of phylogenetic trees. *Journal of Computational Chemistry*, 20(8):867–876, 1999.

[67] B. Schieber and U. Vishkin. On Finding Lowest Common Ancestors: Simplifications and Parallelization. *SIAM Journal of Computing*, 17:12–884, 1988.

[68] Carlos Setubal and Joao Meidanis. *Introduction to Computational Molecular Biology*. PWS Publishing, 1997.

[69] Jeong Seop Sim and Kunsoo Park. The consensus string problem for a metric is NP-complete. *J. Discrete Algorithms*, 1(1):111–117, 2003.

[70] T. F. Smith and M. W. Waterman. Identification of common molecular subsequences. *Journal of Molecular Biology*, 147:195–197, 1981.

[71] Esko Ukkonen. On-Line Construction of Suffix Trees. *Algorithmica*, 14(3):249–260, 1995.

[72] Ron Unger and John Moult. Genetic algorithm for 3d protein folding simulations. In *ICGA*, pages 581–588, 1993.

[73] Vladimir Vapnik. *The Nature of Statistical Learning Theory (Information Science and Statistics)*. Springer-Verlag New York Inc., 2001.

[74] Vijay V. Vazirani. *Approximation Algorithms*. Springer-Verlag Berlin and Heidelberg GmbH & Co. K, 2001.

[75] J. Craig Venter. Sequencing the human genome. In *RECOMB*, pages 309–309, 2002.

[76] M. Vidyasagar. *A Theory of Learning and Generalization (Communications and Control Engineering)*. Springer-Verlag Berlin and Heidelberg GmbH & Co. K, 1996.

[77] Lusheng Wang and Tao Jiang, On the Complexity of Multiple Sequence Alignment. *Journal of Computational Biology*, 1(4):337–348, 1994.

[78] M.S. Waterman. *Introduction to Computational Biology: Maps, Sequences and Genomes (Interdisciplinary Statistics)*. Chapman & Hall/CRC, 1995.

[79] Peter Weiner. Linear pattern matching algorithms. In *FOCS*, pages 1–11, 1973.

[80] Yong L. Xiao and Donald E. Williams. Molecular docking using genetic algorithms. In *SAC '94: Proceedings of the 1994 ACM symposium on Applied computing*, pages 196–200, New York, NY, USA, 1994. ACM.

[81] M. Zuker and P. Stiegler. Optimal computer folding of large RNA sequences using thermodynamics and auxiliary information. *Nucleic Acids Res.*, 9(1):133–148, 1981.

Index